Advances in Biomedical Science and Engineering

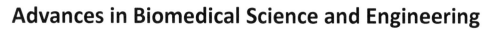

Advances in Biomedical Science and Engineering

Edited by **Mark Walters**

CLANRYE INTERNATIONAL

New Jersey

Published by Clanrye International,
55 Van Reypen Street,
Jersey City, NJ 07306, USA
www.clanryeinternational.com

Advances in Biomedical Science and Engineering
Edited by Mark Walters

© 2015 Clanrye International

International Standard Book Number: 978-1-63240-046-8 (Hardback)

Printed in the United States of America.

Contents

Preface

This book provides a comprehensive account of the advances in biomedical science and engineering. Biomedical science is a vast field dealing with disease progression, paradigms and therapeutic measures. For instance, efficacy of red palm oil and phytomedicines to fight HIV and diabetes by increasing antioxidant activities is analyzed under biomedical sciences. Mathematical drafting of physiological systems and their assessment comes under physiological engineering. Even in hospital management, both biomedical science and engineering are required in order to run hospitals efficiently. This book will benefit students and experts to enhance their knowledge in the above stated topics.

This book is a result of research of several months to collate the most relevant data in the field.

When I was approached with the idea of this book and the proposal to edit it, I was overwhelmed. It gave me an opportunity to reach out to all those who share a common interest with me in this field. I had 3 main parameters for editing this text:

1. Accuracy – The data and information provided in this book should be up-to-date and valuable to the readers.

2. Structure – The data must be presented in a structured format for easy understanding and better grasping of the readers.

3. Universal Approach – This book not only targets students but also experts and innovators in the field, thus my aim was to present topics which are of use to all.

Thus, it took me a couple of months to finish the editing of this book.

I would like to make a special mention of my publisher who considered me worthy of this opportunity and also supported me throughout the editing process. I would also like to thank the editing team at the back-end who extended their help whenever required.

Editor

Biomedical Science: Disease Pathways, Models and Treatment Mechanisms

Biomedical Engineering Professional Trail from Anatomy and Physiology to Medicine and Into Hospital Administration: Towards Higher-Order of Translational Medicine and Patient Care

Dhanjoo N. Ghista

Department of Graduate and Continuing Education,
Framingham State University, Framingham, Massachusetts,
USA

1. Introduction

1.1 Theme

For Biomedical Engineering (BME) to be a professional discipline, we need to address the professional needs of anatomy and physiology, medicine and surgery, hospital performance and management.

The role of BME in Anatomy is to demonstrate how anatomical structures are intrinsically designed as optimal structures. In Physiology, the BME formulation of physiological systems functions can enable us to characterize and differentiate normal systems from dysfunctional and diseased systems. In order to address Medical needs, we need to cater to the functions and disorders of organ systems, such as the heart, lungs, kidneys, and the glucose regulatory system. In Surgery, we can develop the criteria for candidacy for surgery, carry out pre-surgical analysis of optimal surgical approaches, and design surgical technology and implants. In Hospital management, we can develop measures of cost-effectiveness of hospital departments, budget development and allocation.

For BME in Anatomy, we depict how the spinal disc is designed as an intrinsically optimal structure. For BME in Medicine, we formulate the engineering system analyses of Organ system functions and medical tests,

- in the form of differential equations (DEqs), expressing the response of the organ system in terms of monitored data,
- in which the parameters of the DEq are selected to be the organ system's intrinsic functional performance features.

The solution of the organ system's governing DEq is then derived, and made to simulate the monitored data, in order to:

- evaluate the system parameters,
- and obtain the normal and disease ranges of these parameters.

These parameters can then be combined into a Non-dimensional Physiological Index (NDPI),

- by which the system can be assessed in terms of just one "number",
- whose normal and disease ranges can enable effective medical assessment.

Herein, we demonstrate how this methodology [1, 2] is applied to:

- Treadmill test, for evaluating cardiac fitness;
- Lung ventilation modelling, for assessment of status of mechanically ventilated COPD patients;
- Derive a Cardiac Contractility index, which can be determined non-invasively (in terms of auscultatory pressures) and applied to assess left ventricular contractile capacity;
- Glucose Tolerance tests, to detect diabetic patients and borderline patients at risk of becoming diabetic;
- Non-invasive determination of Aortic Pressure profile, systemic resistance and aortic elastance (*Eao*, to characterize the LV systemic load).

Finally, we have also shown the application of this concept and methodology to hospital management. There is a considerable (and hitherto under developed) scope for application of Industrial Engineering discipline for effective hospital administration, in the form of how to determine and allocate hospital budget to optimise the functional performances of all the hospital departments. This leads us to what can be termed as the *Hospital Management System*.

Herein, we have shown how to formulate a performance index (PFI) for ICU. This index divided by the Resource index gives us the cost-effectiveness index (CEI). The Management strategy is to maintain certain acceptable values of both PFI and CEI for all hospital departments, by judicious allocation of staff to the departments. This enables the determination of the Optimal Resource index (RSI) and hospital budget (HOB) to maintain a balance between PFI and CEI for all the hospital departments. This can constitute the basis of Hospital Management.

2. Anatomy: Spine analysed as an intrinsically optimal structure

2.1 Spinal vertebral body (an intrinsically efficient load-bearer)

The spinal vertebral body (VB) geometry resembles a hyperboloid (HP) shell (fig 2) which is loaded by compressive and torsional loadings, portrayed in fig 1 as resolved into component forces along its generators.

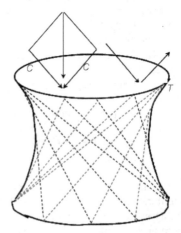

Fig. 1. Vertebral body, a hyperboloid (HP) shell formed of 2 sets of generators [3].

Stress analysis of the VB under Axial Compression [3]:

We now analyse for stresses in the HP shell (generators) due to a vertical compressive force P, as shown in figures 3 and 4. Assume that there are two sets of 'n' number of straight bars placed at equal spacing of $(2\pi a/n)$ measured at the waist circle, to constitute the HP surface, as shown in figure 3 (right). Due to the axi-symmetric nature of the vertical load, no shear stresses are incurred in the shell, i.e. $\sigma_{\phi\theta}=0$, as in figure 3 (left). We then delineate a segment of the HP shell, and consider its force equilibrium (as illustrated in figure 4), to obtain the expressions for stresses N_ϕ and N_θ as depicted in figure 4.

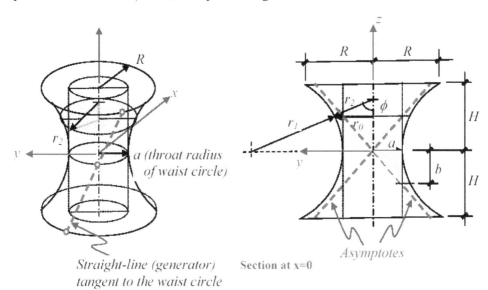

Straight-line (generator)   Section at x=0
tangent to the waist circle

r_1: radius of curvature of meridian

r_2: slanted radius of horizontal section having radius r_o

Kinematic Relationship:

$$r_o = r_2 \,(sin\phi)$$

$$r_1 = -\left(\frac{b^2}{a^4}\right)r_2^3$$

Equation of HP curves:

$$\frac{x^2 + y^2}{a^2} - \frac{z^2}{b^2} = 1$$

At $x = 0$, $\dfrac{y^2}{a^2} - \dfrac{z^2}{b^2} = 1$

Equation of asymptotes:

$$z = \pm\left(\frac{b}{a}\right)y$$

Fig. 2. Geometry of a Hyperboloid (HP) shell. In the figure z = b, and y = a. We define tan β = a/b [3].

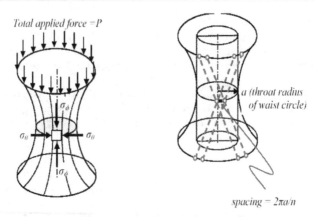

Fig. 3. Stress Analysis for Vertical Loading: Stresses at the waist section of the VB HP Shell: (a) stress components (b) equivalent straight bars aligned with the generators) placed at equal spacing to take up the stresses. In fig 3 (left) due to axi-symmetric vertical load, no shear stresses are incurred in the shell, i.e. $\sigma_{\phi\theta}=0$. In fig 3 (b), there are 2 sets of 'n' number of straight bars placed at equal spacing of $(2\pi a/n)$ measured at the waist circle, to constitute the generators of the HP surface [3].

Equilibrium of Forces on a Shell Segment under Vertical load P:

At the waist ($r_0= a$),

$N_\phi = \dfrac{P}{2\pi a}$, compressive

Now since, $\dfrac{N_\phi}{r_1} - \dfrac{N_\theta}{r_2} = p$,

Hence,

$N_\theta = \left(\dfrac{a^2}{b^2}\right)\dfrac{P}{2\pi a} = \dfrac{P\tan^2\beta}{2\pi a}$

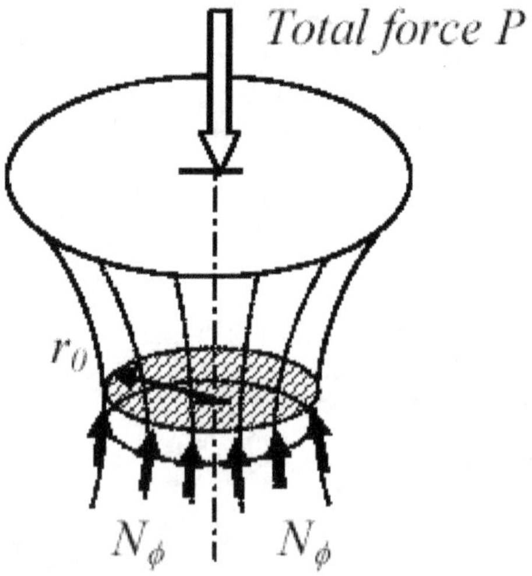

Fig. 4. Equilibrium of Forces on a Shell Segment: Analysis for stresses N_ϕ and N_θ due to the vertical force P [3,2].

Then based on the analysis in Fig 5, we obtain the expression for the equivalent resultant compressive forces C in the fibre-generators of the VB HP shell. Thus it is seen that the total axial loading is transmitted into the HP-shell's straight generators as compressive forces. It is to be noted that the value of C is independent of dimensions R and a.

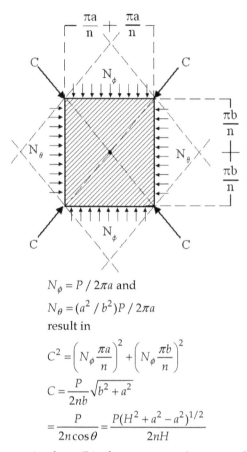

$N_\phi = P / 2\pi a$ and

$N_\theta = (a^2 / b^2) P / 2\pi a$

result in

$$C^2 = \left(N_\phi \frac{\pi a}{n} \right)^2 + \left(N_\phi \frac{\pi b}{n} \right)^2$$

$$C = \frac{P}{2nb} \sqrt{b^2 + a^2}$$

$$= \frac{P}{2n\cos\theta} = \frac{P(H^2 + a^2 - a^2)^{1/2}}{2nH}$$

Fig. 5. Equivalent compressive force C in the generators (corresponding to the stress-resultants acting on the shell element) equilibrating the applied axial loading [3,2].

It can be noted that the value of C is independent of dimensions R and a.

Stress analysis under Torsional loading [3]:

Next, we analyse the compressive and tensile forces in the HP shell generators when the VB is subjected to pure torsion (T).In this case (referring to **fig. 6**), the normal stress resultants are zero, and we and only have the shear stress-resultant , as given by

$$N_\phi = N_\theta = 0 \quad \text{and} \quad N_{\phi\theta} = \tau t$$

The equilibrium of a segment of the shell at a horizontal section at the waist circle (depicted in **figure 6 a**) gives the shear stress-resultant as follows:

$$[(\tau \cdot t)(2\pi a)]a = T \text{ , i.e., } N_{\phi\theta} = \frac{T}{2\pi a^2}$$

Stress Analysis for Torsional Loading M

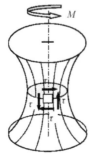

For equilibrium,

$$(2\pi a)(\tau \cdot t)a = M$$

$$\tau = \frac{M}{2\pi a^2 t}$$

$$N_{\phi\theta} = \frac{M}{2\pi a^2}$$

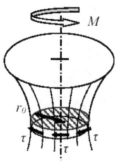

Equilibrium of a shell segment under torsion (M) and shear stresses (τ) (or shear stress-resultant N$_{\phi\theta}$)

Stresses in the HP shell element ($\sigma_\phi=\sigma_\theta=0$ and $\sigma_{\phi\theta}= \tau$) due to torsion M acting on the VB

Fig. 6. Stress analysis of the vertebral body under torsional loading [3].

Now, we consider an element at the waist circle as shown in **figure** 7. The equivalent compressive force (F_{cT}) and tensile force (F_{tT}), in the directions aligned to their respective set of shell generators, are given by

$$F_{cT}^2 = F_{tT}^2 = \left(N_{\phi\theta}\frac{\pi a}{n}\right)^2 + \left(N_{\phi\theta}\frac{\pi b}{n}\right)^2$$

or,

$$\left|F_{cT}\right| = \left|F_{tT}\right| = \frac{T}{2na\sin\beta}$$

wherein F_{cT} and F_{tT} are depicted as c and T respectively in figure 7.

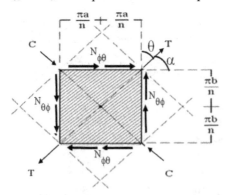

$$= \frac{M(a^2 + a^2\cot^2\theta)^{1/2}}{2na^2}$$

$$= \frac{M}{2na\sin\theta}\left|T\,or\,C\right|$$

$$= \frac{M\cos ec\,\theta}{2na}$$

Fig. 7. Analysis of equilibrium of a shell element comprising of two intersecting generators: Expressions for tensile forces T and compressive forces C in the generators, indicate that torsion loading is also transmitted as axial compressive and tensile forces through the generators of the VB, which makes it a naturally optimum (high-strength and light-weight) structure [3].

Biomedical Engineering Professional Trail from Anatomy and Physiology to Medicine and Into Hospital Administration:
Towards Higher-Order of Translational Medicine and Patient Care

9

Thus, a torsional loading on the VB HP shell is taken up by one set of generators being in compression and the other set of generators being in tension.

Equilibrium of a Shell Element Comprising of Two Intersecting Generators: The equivalent compressive forces (C) and tensile forces (T); in the shell generators (required to equilibrate the applied load), are given by:

$$|T| = |C| = \sqrt{\left(\tau \cdot t\left(\frac{\pi a}{n}\right)\right)^2 + \left(\tau \cdot t\left(\frac{\pi b}{n}\right)\right)^2} = \tau \cdot t\left(\frac{\pi}{n}\right)\sqrt{a^2 + b^2} = \frac{M(a^2 + b^2)^{1/2}}{2na^2}$$

2.2 Spinal disc optimal design (to bear loading with minimal deformation and maximal flexibility)

Fig 8 illustrates how the mechanism of how the spinal disc bears compression without bulging. It is seen that the nucleus pulposus plays the key role in this mechanism, as will be explained further in this section. Hence the absence of it in a denucleated disc causes the disc to bulge.

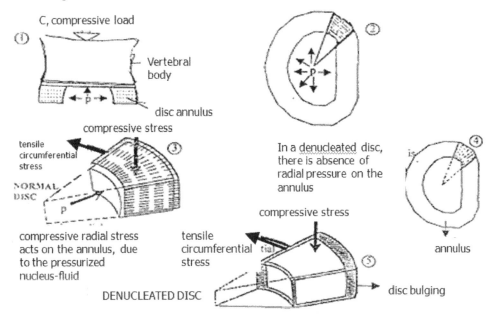

Fig. 8. Mechanism of how the spinal disc bears compression without bulging.

Fig 9 displays the geometry and the deformation variables of the spinal disc. We now present the elasticity analysis of the disc to first obtain the radial, circumferential and axial stresses in terms of the disc deformations and annulus modulus E[4].

We next carry out the stress analysis of the disc under vertical loading P (fig 10), to obtain the expressions (equations 10) for the stresses in the disc annulus in terms of the load P, pressure p in the nucleus-pulposus, and the disc dimensions (a and b) [4].

We then derive the expressions for the disc axial and radial deformations δ_u and δ_n in terms of nucleus pulposus pressure p, the annulus modulus E and the disc dimensions, as given by equations (17) and (18). Now the annulus modulus E is a function of the stress in the annulus (k being the constitutive proportionality constant), and hence of the pressure p and the disc dimensions, as shown by equation (21). As a result of this relationship, we finally show that the disc deformations are only functions of k and the disc dimensions. This implies that irrespective of the increase in the value of load P, the disc deformations remain constant, and only depend on the constitutive property parameter k. This is the novelty of the intrinsic design of the spinal disc!

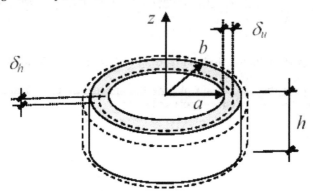

Fig. 9. Geometry and Deformation Variables of the Spinal Disc [4].

ANALYSIS

The equilibrium equation is:

$$\sigma_r - \sigma_\theta + r\frac{d\theta_r}{dr} = 0 \tag{1}$$

Strain-displacement relations:
$$\epsilon_r = \frac{du}{dr}, \epsilon_\theta = \frac{u}{r}, \tag{2}$$

Constitutive relations:

$$\sigma_r = E\epsilon_r = E\frac{du}{dr}, \quad \sigma_\theta = E\epsilon_\theta = E\frac{u}{r} \tag{3}$$

Substituting eqn (3) into eqn (1), we have:

$$\frac{d^2u}{dr^2} + \frac{1}{r}\frac{du}{dr} - \frac{u}{r^2} = 0, \quad u = Ar + \frac{B}{r} \tag{4}$$

Because of the incompressible nucleus pulposus fluid inside:

$$\pi a^2 h = \pi(a + \delta_u)^2(h - \delta_h) \tag{5}$$

the (radial and axial) deformations δ_u and δ_h are related as:

$$\delta_u = \frac{\pi a^2 \delta_h}{2\pi ah} = \left(\frac{a}{2h}\right)\delta_h \tag{6}$$

Now, we designate:

$$u\big|_{r=a} = \delta_u, \sigma_r\big|_{r=b} = 0 \tag{7}$$

Substituting eqn (7) into eqns (3 & 4), we get

$$Aa + \frac{B}{a} = \delta_u$$

$$E\left[A - \frac{B}{b^2}\right] = 0 \tag{8}$$

So that

$$A = \frac{a\delta_u}{a^2 + b^2}, \quad B = \frac{a\delta_u \cdot b^2}{a^2 + b^2} \quad \text{in the 'u' function} \tag{9}$$

Then, substituting A and B into eqns (4 & 3), we obtain:

- for the radial stress,

$$\sigma_r = \frac{Ea\left(r^2 - b^2\right)\delta_u}{(a^2 + b^2)r^2} = \frac{a^2\left(r^2 - b^2\right)}{2h(a^2 + b^2)r^2}E\delta_h \tag{10-a}$$

- for the circumferential stress,

$$\sigma_\theta = \frac{Ea\left(r^2 + b^2\right)\delta_u}{(a^2 + b^2)r} = \frac{a^2\left(r^2 + b^2\right)}{2h(a^2 + b^2)r}E\delta_h \tag{10-b}$$

- for the axial stress

$$\sigma_z = E\epsilon_z = E\frac{\delta_h}{h} \qquad \delta_u = \left(\frac{a}{2h}\right)\delta_h \tag{10-c}$$

Once σ_z is evaluated, δ_h will become known (from eqn. 10-c) and subsequently σ_r, σ_θ (from eqns 10-a & 10-b) and δ_u (from eqn 10-c).

Stress Analysis for Vertical Loading [4]:

For a vertically applied force P,

$$P = \left(\pi a^2\right)\sigma_f + \pi(b^2 - a^2)\sigma_z \tag{11}$$

where σ_f is the hydrostatic pressure in the NP fluid, and σ_z the axial stresses in the annulus. Because the disc height (h) is small, $\sigma_f \approx constant$, and hence

$$\sigma_f = -\sigma_r\big|_{r=a} = p \tag{12}$$

Then, based on eqns (12, 10-a & 10-c), we obtain

$$\sigma_f = -\frac{\left(a^2 - b^2\right)}{2h\left(a^2 + b^2\right)}E\delta_h = \frac{\left(b^2 - a^2\right)}{2h\left(a^2 + b^2\right)}E\left(\frac{\delta_h}{h}\right), \text{ or } \sigma_f = \frac{(b^2 - a^2)}{2(a^2 + b^2)}\sigma_z = p \tag{13}$$

Note:

σ_f : induced fluid pressure

σ_z : compressive stress induced in the annulus

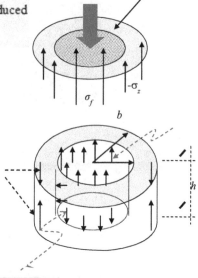

Normal stresses σ_f & σ_z equilibrating the applied force P

Fig. 10. Induced Stresses in the disc annulus and Pressure p in the nucleus-pulposus in response to the load P [4].

Substituting

$$\sigma_f = \frac{\left(b^2 - a^2\right)}{2\left(a^2 + b^2\right)}, \quad \sigma_z = p, \quad \sigma_z = E\frac{\delta_h}{h} = 2E\frac{\delta_u}{a}$$

Into

$$P = (\pi a^2)\sigma_f + \pi(b^2 - a^2)\sigma_z$$

We get:

$$\delta_h = \frac{h\sigma_z}{E} = \frac{h}{E} \frac{P}{\pi(b^2 - a^2)}\left[\frac{2(a^2 + b^2)}{3a^2 + 2b^2}\right] \tag{14}$$

$$\delta_u = \frac{a}{2h}\delta_h = \frac{a}{2E} \frac{P}{\pi(b^2 - a^2)}\left[\frac{2(a^2 + b^2)}{3a^2 + 2b^2}\right] \tag{15}$$

Nucleus-pulposus fluid pressure, $\quad p = \frac{P}{\pi(3a^2 + 2b^2)} \tag{16}$

$$\sigma_z = \frac{2p(a^2 + b^2)}{(b^2 - a^2)} = \frac{2P(a^2 + b^2)}{\pi(b^2 - a^2)(3a^2 + 2b^2)} \tag{16-a}$$

$$\delta_h = \frac{h\sigma_z}{E} = \frac{h}{E}\frac{P}{\pi(b^2 - a^2)}\left[\frac{2(a^2 + b^2)}{3a^2 + 2b^2}\right]$$

Stress in the Annulus

From,

$$\sigma_\theta = \frac{Ea\left(r^2 + b^2\right)\delta u}{(a^2 + b^2)r} = \frac{a^2\left(r^2 + b^2\right)}{2h(a^2 + b^2)r}E\delta h$$

We get :

$$\frac{a^2\left(r^2 + b^2\right)P}{\pi r^2(b^2 - a^2)(3a^2 + 2b^2)} = \frac{pa^2\left(r^2 + b^2\right)}{r^2(b^2 - a^2)}, \tag{17}$$

$$\sigma_\theta(r = a) = \frac{p\left(a^2 + b^2\right)}{(b^2 - a^2)}, \text{ in the annulus} \tag{18}$$

wherein, $p = \dfrac{P}{\pi\left(3a^2 + 2b^2\right)}$, pressure in nucleus-pulposus fluid

The disc deformations have been obtained as:

Axial deformation,

$$\delta_h = \frac{2ph(a^2 + b^2)}{E_c(b^2 - a^2)} \tag{19}$$

Radial deformation

$$\delta_u = \frac{pa(a^2 + b^2)}{E_c(b^2 - a^2)} \tag{20}$$

wherein $E_c = E - E_o = k\sigma$.

$$\therefore E_c = k\sigma_\theta(r = a) = kp\frac{(a^2 + b^2)}{(b^2 - a^2)} \tag{21}$$

Now, as the magnitude of the load P increases, the pressure p in nucleus-pulposus fluid also increases. Then, as p increases, so does the modulus E_c according to eqn (21)

$$\therefore \frac{p}{E_c} = \frac{(b^2 - a^2)}{k\left(a^2 + b^2\right)} = \text{constant} \tag{23}$$

$$\delta_h = \frac{2ph(a^2 + b^2)}{E_c(b^2 - a^2)} = \frac{2h}{k} \text{ , a constant, and } \delta_u = \frac{pa(a^2 + b^2)}{E_c(b^2 - a^2)} = \frac{2a}{k} \text{ , a constant}$$

This means that, irrespective of increase in the value of load P, the disc deformations δ_u and δ_h remain constant, and only depend on the constitutive property parameter k. This is the novelty of the intrinsic design of the spinal disc!

3. Physiology: Mechanism of left ventricle twisting and pressure increase during isovolumic contraction (due to the contraction of the myocardial fibres)

Introduction and objective: The left ventricular (LV) myocardial wall is made up of helically oriented fibers. As the bioelectrical wave propagates along these fibers, it causes concomitant contraction wave propagation. Our LV cylindrical model is illustrated in figure 11. The contraction of the helical oriented myocardial fibers causes active twisting and compression of the LV (as illustrated in fig 11), thereby compressing the blood fluid contained in it. Then due to the very high bulk modulus of blood, this fluid compression results in substantial pressure increase in the LV cavity.

Herein we simulate this phenomenon of LV isovolumic contraction, which causes the intra-LV pressure to rise so rapidly during 0.02-0.06 seconds of isovolumic contraction. Our objective is to determine how the pressure generated during isovolumic contraction, due to by active torsion (with LV twisting) and compression (with LV shortening) caused by the contractile stress in the helically wound myocardial fibers [5].

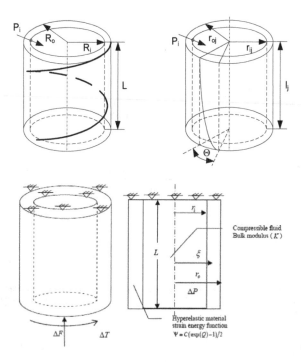

Fig. 11. Top: Fiber orientation and twisting model of the left ventricle (LV). Bottom: The fluid-filled LV cylindrical shell model: (i) geometry (ii) material property, and (iii) equivalent compression ΔF and ΔT associated with its internal stress state due to internal pressure rise within it.

Concept: In order to simulate the left ventricle deformation during isovolumic contraction, we have modeled it as a pressurized fluid-filled thick-walled cylindrical shell supported by the aorta along its upper edge. The LV cylindrical model consists of an incompressible hyperelastic material with an exponential form of the strain energy function ψ.

The contraction of the myocardial fibers causes active twisting and compression of the left ventricle, thereby compressing the blood fluid contained in it. Then, due to the very high bulk modulus of blood, this compression results in pressure increase in the ventricular cavity. Hence, we simulate this phenomenon by applying and determining equivalent active torque and compression to the LV-cylindrical model incrementally (as ΔF and ΔT), so as to raise the LV pressure by the monitored amounts.

Modeling approach:

We monitor LV pressure (p), LV volume (V), myocardial volume (V_M), and wall thickness (h) at time intervals during isovolumic contraction. From the monitored V, V_M, and h, we determine the LV model radius R and length L and the wall thickness h. We also monitor LV twist angle ϕ.

We then invoke blood compressibility to determine ΔV at subsequent instants , as $\Delta V = (\Delta p/K) V$, where Δp is the monitored incremental pressure and K is the bulk modulus of blood. From the volume strain $\Delta V/V$, we then determine the model length and radius strains $\Delta I/L$ and $\Delta r/R$, and hence the LV dimensions with respect to LV dimensions at the start of isovolumic contraction. This enables us to determine the stretches (strains) $(\lambda_z, \lambda_r, \lambda_\theta$ and $\gamma)$, and thereafter the Lagrange strain tensor components $(E_{rr}, E_{\theta\theta}, E_{zz}, E_{\theta z})$ in terms of these stretches and the hydrostatic pressure.

Then, we express the LV wall stresses in terms of the strain energy density function Ψ, of the Lagrange strain tensor components in cylindrical coordinates and the material constitutive parameters (b_i), in which (i) the stretches $(\lambda_z, \lambda_r, \lambda_\theta$ and $\gamma)$ have been calculated and are known, (ii) the hydrostatic pressure and the constitutive parameters b_i $(i = 1, 2, \ldots , 9)$ are the unknowns.

So now, we substitute the stress expressions σ_{rr} and $\sigma_{\theta\theta}$ into the boundary conditions equations (of equilibrium between the internal pressure and the wall stresses σ_{rr} and $\sigma_{\theta\theta}$, and between the internal pressure and wall stress σ_{zz}), and determine the best values of the constitutive parameters (b_i) and the hydrostatic pressure to satisfy these equations.

We then go back, and determine the stress expressions. We utilize the stress expressions for σ_{zz} and $\sigma_{\theta z}$, to determine the generated values of torsion (ΔT) and axial compression (ΔF),due to the contraction of the helically wound myocardial fibres.

Finally, we determine the principal stresses and principal angle along the radial coordinate of the LV wall thickness, from which we can interpret the fibre orientations, which can be related to the LV contractility index.

This procedure is carried out at three instants of time from the start of isovolumic contraction, and at 0.02 s, 0.04 s, 0.06 s into the isovolumic contraction phase. Hence, from the monitored LV Δp and computed ΔV at these three instants (with respect to the pressure and volume at t = 0 at the start of isovolumic contraction phase), we determine (i) the time variation of the internally generated torque and axial compression during the isovolumic contraction phase (fig 12), as well as (ii) the time variations of the principal (tensile) stress and the principal angle (taken to be equivalent to the fiber angle) during the isovolumic contraction phase (fig 13).

Model Kinematics:

We model the LV as an incompressible thick-walled cylindrical shell subject to active torsion torque and compression as illustrated in fig 11. The upper end of the LV model is constrained in the long-axial direction to represent the suspension of the left ventricle by the aorta at the base. Now, considering the LV at end-diastole to be in the unloaded reference configuration, the cylindrical model in its undeformed state is represented geometrically in terms of cylindrical coordinates (R, Θ, Z) by

$$R_i \leq R \leq R_o, \ 0 \leq \Theta \leq 2\pi, \ 0 \leq Z \leq L \tag{1}$$

where R_i, R_o and L denote the inner and outer radii, and the length of the undeformed cylinder, respectively.

In terms of cylindrical polar coordinates (r, θ, z), the geometry of the deformed LV configuration (with respect to its undeformed state at the previous instant) is given by:

$$r_i \leq r \leq r_o, \ 0 \leq \theta \leq 2\pi, \ 0 \leq z \leq l \tag{2}$$

where r_i, r_o and l denote the inner and outer radii, and the length of the deformed cylinder, respectively.

We further consider the incompressible LV model in its reference state to be subjected to twisting, radial and axial deformations in the radial and long-axis directions during isovolumic contraction, such that (also based on incompressibility criterion), the deformations of incompressible LV cylindrical shell can be expressed as

$$r = \sqrt{\frac{R^2 - R_i^2}{\lambda_z} + r_i^2}, \ \theta = \Theta + Z\frac{\phi}{L}, \ z = \lambda_z Z \tag{3}$$

where λ_z is the constant axial stretch, r_i is the inner radius in the deformed configuration and ϕ is the measured angle of twist at the apex of the LV (relative to the base). It can be seen that the twist angle (θ) and the axial deformation (z) are zero at the upper end of the LV.

Model Dimensions:

At any instant (t), the geometrical parameters (or dimensions) of the LV cylindrical model (instantaneous radius R and length L, as defined in fig 11) can be determined in terms of the monitored LV volume (V), myocardial volume (V_M) and wall thickness (h), as follows:

$$R_i = \frac{2Vh / V_M + \sqrt{(2Vh / V_M)^2 + 4Vh^2 / V_M}}{2} \tag{4-a}$$

$$L = V / \pi R_i^2 \tag{4-b}$$

Then,

$$R_o = R_i + h. \tag{4-c}$$

These equations will be employed to determine the LV dimensions at the start of isovolumic contraction phase (t = 0). The determination of the dimensions of the deformed LV (due to contraction of the myocardial fibers) at the subsequent instants of the isovolumic contraction phase is indicated in the next subsection. We also utilize the information on the LV twist angle (ϕ) during the isovolumic phase, from MRI myocardial tagging. From this information, we can determine the stretches $(\lambda_z, \lambda_r, \lambda_\theta$ and $\gamma)$.

Theoretical Analysis:

The strain energy density function suitable for the myocardium material, is given by:

$$\psi = C(\exp(Q) - 1)/2 \tag{5}$$

wherein Q is a quadratic function of the 3 principal strain-components (to describe 3-d transverse isotropy) in the cylindrical coordinate system, given by:

$$Q = b_1 E_{\theta\theta}^2 + b_2 E_{ZZ}^2 + b_3 E_{RR}^2 + 2b_4 E_{\theta\theta}E_{ZZ} + 2b_5 E_{RR}E_{ZZ}$$
$$+ 2b_6 E_{\theta\theta}E_{RR} + 2b_7 E_{\theta Z}^2 + 2b_8 E_{RZ}^2 + 2b_9 E_{\theta R}^2 \tag{6}$$

wherein b_i are non-dimensional material parameters, and E_{ij} are the components of the modified Green-Lagrange strain tensor in cylindrical coordinates (R, Θ, Z). In order to reduce the mathematical complexity of the problem, we assume negligible shear during isovolumic contraction. Thus E_{RZ} and $E_{\theta R}$ in equation (6) and their corresponding stress components (σ_{RZ} and $\sigma_{\theta R}$) are neglected.

The stress equilibrium equation (in the cylindrical coordinate system) is given by the following equation:

$$\frac{d\sigma_{rr}}{dr} + \frac{(\sigma_{rr} - \sigma_{\theta\theta})}{r} = 0 \tag{7}$$

The boundary conditions on the outer and inner surfaces of the LV cylindrical model are given by

$$\sigma_{rr}(r = r_o) = 0, \quad \sigma_{rr}(r = r_i) = -p \tag{8}$$

where p is the LV pressure acting on the inner surface of the LV model; we employ incremental pressure Δp with respect to the LV pressure at t = 0, the start of isovolumic contraction.

By integrating eq (7), the Cauchy radial stress σ_{rr} is given by:

$$\sigma_{rr}(\xi) = \int_\xi^{r_o} (\sigma_{rr} - \sigma_{\theta\theta})\frac{dr}{r}, \quad r_i \le \xi \le r_o \tag{9}$$

There from, the boundary condition eq (8) of the internal pressure p = -σ_{rr} (r=r$_i$) is obtained (by substituting equation 9 into the boundary condition 8), in the form:

$$p_i = -\int_{r_i}^{r_o} (\sigma_{rr} - \sigma_{\theta\theta})\frac{dr}{r}, \tag{10}$$

Since the valves are closed during isovolumic contraction, we impose another set of boundary conditions (at both the top and bottom of the internal LV surface), giving:

$$\sigma_{zz}\pi\left(r_o^2 - r_i^2\right) = p\left(\pi r_i^2\right) \tag{11}$$

where σ_{zz} denotes the axial component of the Cauchy stresses. In the analysis, we will employ Δp with respect to the pressure at t = 0, at the start of isovolumic contraction.

The blood in the left cavity is assumed compressible, and the change in cavity volume (ΔV) due to the monitored incremental pressure (Δp), is given by

$$\Delta V = \Delta p / K; \qquad K = 2.0 \times 10^9 \, pa \tag{12}$$

where K is the bulk modulus of blood.

Analysis and computational procedure:

The following analysis is carried out at the three time instants t (or j) = 0.02 secs, 0.04 secs and 0.06 secs (from the start t = 0 of the isovolumic contraction phase, from the monitored Δp and computed ΔV (eq 12) at the three time instants with respect to p and V at t = 0 (the start of isovolumic contraction phase).

1. We obtain the increments in LV pressure (Δp) from the monitored pressure data, during isovolumic contraction. By taking $K=\Delta p/\Delta V$, we compute the corresponding changes in LV volume ΔV.
2. Left ventricular deformation: For this change ΔV in the LV volume, by assuming the ratios of cylindrical-model length and radius strains $\Delta l/L$ and $\Delta r/R$ to be equal during isovolumic contraction, we obtain their expressions as:

$$\Delta l_j = \left(1 - \sqrt[3]{1 - \Delta p/K}\right)L, \ \ \Delta r_{ij} = \left(1 - \sqrt[3]{1 - \Delta p/K}\right)R \tag{13}$$

From equation (13), the incremental quantities Δl_j and Δr_{ij} can be calculated, and hence:

$$l_j = L - \Delta l_j, \text{ and } r_{ij} = R - \Delta r_{ij} \tag{14}$$

where l_j and r_{ij} are the deformed model length and radius at time t (or j)
So the wall-thickness h can be obtained from:

$$h_j = \sqrt{\frac{V_M / l_j + \pi r_{ij}^2}{\pi}} - r_{ij} \tag{15}$$

Let $\Delta\phi$ denote the relative angle of twist measured at the apex, at each of the 3 stages of isovolumic contraction phase, obtained by magnetic resonance imaging (MRI).

3. Next we determine the stretches in the 3 directions (due to deformed dimensions l and r with respect to the undeformed dimensions L and R) as follows

$$\lambda_z(r) = \frac{1}{L}, \ \lambda_r(r) = \frac{\partial r}{\partial R} = \frac{R}{r\lambda_z}, \ \lambda_\theta(r) = \frac{r\partial\theta}{R\partial\Theta} = \frac{r}{R} \tag{16}$$

We define the twist stretch due to torsion, as

$$\gamma(r) = \frac{r\partial\theta}{\partial z} = \frac{r\phi}{l} \tag{17}$$

wherein ϕ is zero at the top surface of the LV held by the aorta

4. Next we express the components of the Lagrange Green's strain tensor in terms of the stretches and deformations obtained from equations (8-10), as:

$$E_{rr}=\frac{1}{2}\left(\lambda_r^2-1\right),\ E_{\theta\theta}=\frac{1}{2}\left(\lambda_\theta^2-1\right),\ E_{zz}=\frac{1}{2}\left(\lambda_z^2\left(1+\gamma^2\right)-1\right),\ E_{\theta z}=\frac{\gamma\lambda_z\lambda_\theta}{2} \tag{18}$$

5. Then, by using the strain energy function (eqs 5 and 6), we obtain the expressions of Cauchy stress tensor in terms of the parameters b_i:

$$\sigma_{\theta\theta}=\lambda_\theta^2\frac{\partial\Psi}{\partial E_{\theta\theta}}+2\gamma\lambda_z\lambda_\theta\frac{\partial\Psi}{\partial E_{\theta z}}+\gamma^2\lambda_z^2\frac{\partial\Psi}{\partial E_{zz}}-\bar{p}\quad \sigma_{rr}=\lambda_r^2\frac{\partial\Psi}{\partial E_{rr}}-\bar{p}$$

$$\sigma_{zz}=\lambda_z^2\frac{\partial\Psi}{\partial E_{zz}}-\bar{p}\quad \sigma_{\theta z}=\lambda_z\lambda_\theta\frac{\partial\Psi}{\partial E_{\theta z}}+\gamma\lambda_z^2\frac{\partial\Psi}{\partial E_{zz}} \tag{19}$$

wherein ψ is given by equations (5 & 6), and the to-be-determined unknown parameters are the hydrostatic pressure $\bar{p}$ and the constitutive parameters b_i (in equation 6).

6. We now employ the stress expressions into the boundary conditions (10 & 11), and determine the hydrostatic pressure $\bar{p}$ and the parameters b_i (i=1,2,...7) and, using a nonlinear least squares algorithm.

7. We next determine (i) the stretches from eqs (16 & 17), and (ii) therefrom, the strains $E_{i,j}$ from eq (18). Then by substituting the computed strains components $E_{i,j}$ and the strain energy density function ψ (eqs 5 & 6) into eq (19), we can determine the LV wall stress components for the LV cylindrical model.

8. From the stress components $\sigma_{\theta z}$, $\sigma_{\theta\theta}$ and σ_{zz}, we can computes the principal stresses σ_1, σ_2 the principal angle ϕ.

$$\sigma_{1,2}=\frac{\sigma_{zz}+\sigma_{\theta\theta}}{2}\pm\sqrt{\left(\frac{\sigma_{zz}-\sigma_{\theta\theta}}{2}\right)^2+\sigma_{\theta z}^2} \tag{20}$$

$$\tan2\phi=\frac{2\sigma_{\theta z}}{\sigma_{\theta\theta}-\sigma_{zz}} \tag{21}$$

9. Then, we compute the equivalent active axial force ΔF and the torsional couple ΔT as follows:

$$\Delta F=2\pi\int_{r_i}^{r_o}\sigma_{zz}rdr,\quad M_T=2\pi\int_{r_i}^{r_o}\sigma_{\theta z}r^2dr, \tag{22}$$

Results and comments:

1. The computational model results for a sample subject are provided in Tables 1 -5.

2. The variations of the equivalent active torque T and axial compressive force F during the isovolumic phase are calculated and shown in Fig. 12.

We hence demonstrate that the big increment of internal pressure in LV cavity is caused by the compression of the blood (due to its high bulk modulus) by the internally generated torque and axial force during the isovolumic phase, due to the contraction of the myocardial fibers.

3. The radial variations of the principal stress and principal angle are found to be almost uniform across the wall thickness [5]. The time variation of the principal stress and angle during isovolumic phase are shown in **Fig. 13**. At the end of the isovolumic phase, the magnitude of the tensile principal stress is around 1.75×10^5pa, which is in good agreement with the isometric tension value of 1.40×10^5pa achieved under maximal activation.

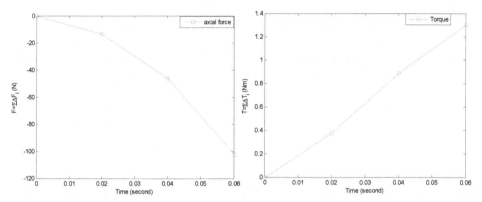

Fig. 12. Variations of axial force and torque as function of time during isovolumic phase [5].

The notable result from **Fig. 13** is that both the principle stresses and their orientation angle keep changing during the isovolumic phase. It is seen from Figure 13, that the myocardial fiber was oriented 38^0 at the start and became 33^0 at the end of the isovolumic phase, as the monitored internal pressure increased during isovolumic phase (due to active torsion and compression, corresponding to active contraction of the helically woven fibers from 38^0 to 33^0) from 24 to 44 to 62 mmHg during the isovolumic contraction phase. Observing the values of the principle stresses, it is seen that the LV is acted upon by internally generated (i) active torsion (T) causing its twisting, and (ii) compression (F) to cause compression of the blood and develop the LV pressure rise.

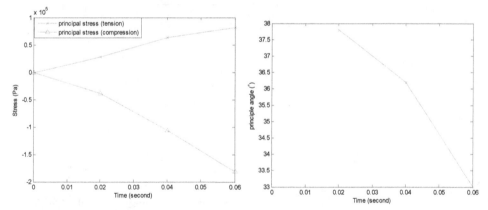

Fig. 13.Variation of principal stresses as functions of time, during isovolumic phase [5].

In conclusion: we have indirectly shown that the contraction of the myocardial fibers (i) develops active stresses in the LV wall (and corresponding principal stresses), and (ii) causes LV twist and shortening, resulting in the development of substantial pressure increase during isovolumic contraction.

The tensile principal stress corresponds to the active contractile stress generated in the myocardial fibers, while the angle of the tensile principal stress corresponds to the fiber helical angle, which is in agreement with the experiment data on the fiber angle.

The computational results are shown in Tables 1 to 5, for a sample subject.

Biomedical Engineering Professional Trail from Anatomy and Physiology to Medicine and Into Hospital Administration: Towards Higher-Order of Translational Medicine and Patient Care

21

t (second)	P (mmHg)	ΔP (mmHg)	V (ml)	ΔV (ml)	r_{ij} (cm) (=R_i)	Δr_{ij} (cm)	l_j (cm) (=L)	Δl_j (cm)	h_j (cm)	r_{oj} (cm) (R_o)	ΔΦ (°)
0	18		1.36700000E+02		2.03208400E+00		1.0537450000E+01		1.0852470000E+00	3.1173310000E+00	0
0.02	43	25	1.36699773E+02	2.27263750E-04	2.032083992E+00	8.46701669E-09	1.053744996E+01	4.39060418E-08	1.085246679E+00	3.117330670E+00	0.667
0.04	63	45	1.366995591E+02	4.09074750E-04	2.032083985E+00	1.52406301E-08	1.053744992E+01	7.90308757E-08	1.085246684E+00	3.117330669E+00	1.333
0.06	81	63	1.366994427E+02	5.72704650E-04	2.032083979E+00	2.13368822E-08	1.053744989E+01	1.10643226E-07	1.085246689E+00	3.117330667E+00	2.00

Table 1. Pressure-volume and model parameters for a sample subject with V_M=185ml.

parameter	b_1	b_2	b_3	b_4	b_5	b_6	b_7	b_8	b_9
value	5946.2278	15690.58158	422.514993	16157.10454	16360.53744	33299.28998	680.7385218	0	0

Table 2. The parameters of the strain energy function for the sample case shown in Table 1.

t	λ_z	λ_r	λ_θ	Y
0				
0.02	0.9999999958	1.000000008E+00	9.99999996E-01	1.28626948E-01
0.04	0.9999999925	1.000000015E+00	9.99999993E-01	2.56482520E-01
0.06	0.9999999895	1.000000021E+00	9.99999990E-01	3.85688000E-01

Table 3. The stretches calculated from equations (16 & 17) for the sample case shown in Table 1.

t (second)	Endocardium								Epicardium	
	0	1.99E-02	1.02E-01	2.37E-01	4.08E-01	0.591715	0.762765	0.89833	0.98014	1
0.02	-24.0809	-23.4615	-17.2801	-12.9839	-8.73378	-4.9888	-2.11726	-0.41173	-0.2144	0
0.04	-44.2917	-43.1341	-31.6543	-23.7482	-15.9694	-9.13029	-3.88098	-0.75569	-0.3542	0
0.06	-6.21E+01	-6.05E+01	-5.43E+01	-4.48E+01	-3.39E+01	-2.31E+01	-1.34E+01	-5.75E+00	-1.13E+00	0

Table 4. Radial stresses distributions along LV wall from endocardium to epicardium.

t (second)	Results								
	Circumferential stress (Pa)	Radial stress (Pa)	Axial force (N)	Torque (Nm)	Axis stress (Pa)	Shear stress (Pa)	Principal stress-tension (Pa)	Principal stress-compression (Pa)	Principal direction (°)
0	0.00	0.00	0.00	0.00	0.00E+00	0.00E+00	0.00	0.00	
0.02	3358.33	-2844.53	-13.51	0.38	-1.35E+04	3.29E+04	28221.68	-38395.38	37.82
0.04	5318.61	-5956.38	-46.06	0.89	-4.70E+04	8.21E+04	63836.28	-105547.96	36.18
0.06	4481.83	-8530.94	-102.38	1.30	-1.04E+05	1.22E+05	81673.18	-180897.35	33.02

Table 5. Results for the sample subject at different instants

4. Clinical evaluation of Physiological systems in terms of Non-dimensional physiological Indices

Non-dimensional numbers (made up of several phenomenon related terms) are employed to characterise – disturbance phenomena. For example, in a cardiovascular fluid-flow regime, the Reynold'snumber

$$N_{re} = \rho VD / \mu \qquad (1)$$

(*V* : flow velocity, *D*: diameter, μ: fluid viscosity, ρ: fluid density) characterizes turbulent flow, which can occur in the ascending aorta when the aortic valve is stenotic (giving rise to murmurs) and accentuated in the case of anaemia (decreased blood viscosity).

Integration of a number of isolated but related events into one non-dimensional physiological index (NDPI) can help to characterise an abnormal state of a particular physiological system [1].

For utilization of an NDPI diagnostically, we evaluate it in a large patient population and develop its distribution into normal and dysfunctional ranges, as illustrated in figure 14. Then, upon evaluation of the NDPI of a particular subject, we can see if that value falls in the normal or dysfunctional range, and accordingly make the medical diagnosis.

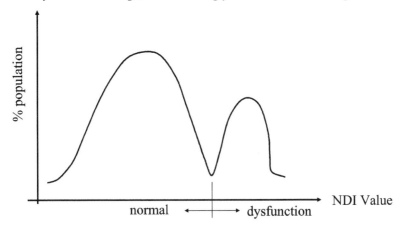

Fig. 14. Illustration of the Distribution of an NDI, showing its normal and dysfunctional ranges which can be employed for diagnostic purpose.

5. Medical test: Cardiac fitness index based on treadmill HR variation

In this procedure, the cardiac fitness model consists of a first-order differential-equation system model, describing the heart rate (HR) response (y) to exertion (exercise, jogging etc) monitored in terms of the work-load, where y is defined as follows:

$$y = \frac{HR(t) - HR(rest)}{HR(rest)} \tag{1}$$

The subject is exercised on the treadmill for a period of time te (minutes) at a constant work load (W), while the HR(t) (and hence y) is monitored, as displayed in fig 15. Now we develop a model to simulate (i) the y(t) response during exercise, i.e. during $t \le te$, and (ii) thereafter for y(t) decay, after the termination of exercise. In a way, te represents the exercise endurance of the subject [6].

For a person, the model equation for HR response is represented by:

$$\frac{dy}{dt} + k_1 y = C_0 W \tag{2}$$

For $t \le t_e$, the solution is given by:

$$y = \frac{y_e(1-e^{-k_1 t})}{(1-e^{-k_1 t_e})} \tag{3}$$

For the recovery period *(t ≥ te)*, the solution of eqn. (2) is :

$$y = y_e e^{-k_2(t-t_e)} \tag{4}$$

where k_1 and k_2, are the model parameters, which can serve as cardiac-fitness parameters (in min^{-1}).

Non-dimensional Cardiac Fitness Index:
A typical y(t) response is illustrated in Fig 15.

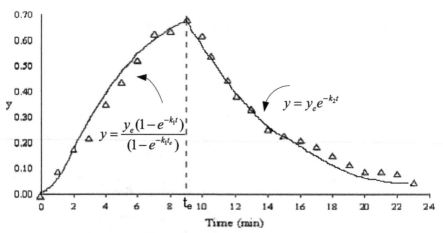

Fig. 15. Graph of y (the computed HR response) vs. t during treadmill exercise *t=te* and during the recover of period *t=te*.

The parameters k_1 and k_2 can be continued into a single nondimensional cardiac-fitness index (CFI):

$$CFI = k_1 k_2 t_e^2 \tag{5}$$

According to this formulation of CFI, a healthier subject has (i) greater k_1 (i.e., slower rate-of-increase of HR during exercise (ii) greater k_2 (i.e., faster rate-of-decrease of HR after exercise) (iii) greater t_e (i.e., exercise endurance), and hence (iv) higher value of CFI.

Subject	Classification	CFI (relative values)
1	Athletic	271
2	Occasionally runs	90
3	Rarely exercise	65
4	Sedentary	19
5	Exercises regularly	155

Table 1. CFI values for athletic, fit and unfit subjects.

Now, we need to evaluate CFI for a big spectrum of patients, and then compute its distribution curve, to determine the efficiency of this index, in order to yield distinct separation of CFI ranges for healthy subjects and unfit patients. This CFI can also be employed to assess improvement in cardiac fitness following cardiac rehabilitation regime. This CFI is non-dimensional, and it can be useful to clinicians as they are able to predict the heart condition or fitness performance of a person by referring to the value of a single index value.

6. Medical physiology: A non-dimensional diabetes index with respect to Oral-Glucose-Tolerance testing

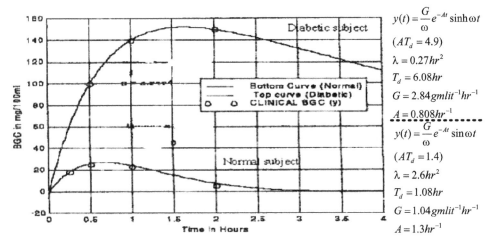

$$y(t) = \frac{G}{\omega} e^{-At} \sinh \omega t$$

$$(AT_d = 4.9)$$

$$\lambda = 0.27 hr^2$$

$$T_d = 6.08 hr$$

$$G = 2.84 gmlit^{-1} hr^{-1}$$

$$A = 0.808 hr^{-1}$$

$$y(t) = \frac{G}{\omega} e^{-At} \sin \omega t$$

$$(AT_d = 1.4)$$

$$\lambda = 2.6 hr^2$$

$$T_d = 1.08 hr$$

$$G = 1.04 gmlit^{-1} hr^{-1}$$

$$A = 1.3 hr^{-1}$$

Fig. 16. OGTT response-Curve: A=1.3hr-1 (i.e., higher damping.coefficient value) for the normal subject for the diabetic patient A=0.808 hr-1, i.e., the damping coefficient is smaller [7, 8].

Oral Glucose Tolerance Test (OGTT) is a standard procedure for diagnosing diabetes and risk of becoming diabetic. However, the test data is assessed empirically. So, in order to make this procedure more reliable, we have carried out a biomedical engineering analysis of the OGTT data, to show how to distinguish diabetes subjects and those at risk of becoming diabetic.

For Oral-glucose Tolerance Test simulation (entailing digestive & blood-pool chambers), the differential equation is as follows [7,8]:

$$y'' + 2Ay' + \omega_n^2 y = G\delta(t); \quad y \text{ in gms/liter, } G \text{ in gm/liter/hr}$$

or,   (1)

$$y'' + \lambda T_d y' + \lambda y = G\delta(t)$$

wherein $\left(\omega_n = \lambda^{1/2} \right)$ is the natural oscillation-frequency of the system, A is the attenuation or damping constant of the system, $\omega = (\omega_n^2 - A^2)^{1/2}$ is the (angular) frequency of damped oscillation of the system, $\lambda = 2A/T_d =, v\omega_n^2$ with (λy) representing the proportional-control term of blood-glucose concentration (y).

($\lambda T_d y'$)is the derivative feedback control term with derivative-time of T_d $G\delta(t)$ represents the injected glucose bolus.

The input to this system is taken to be the impulse function due to the orally ingested glucose bolus [G], while the output of the model is the blood-glucose concentration response y(t). For an impulse glucose-input, a normal patient's blood-glucose concentration data is depicted in **Figure 16** by open circles. Based on the nature of this data, we can simulate it by means of the solution of the Oral-glucose regulatory (second-order system) model, as an under-damped glucose-concentration response curve, given by:

$$y(t) = (G / \omega)e^{-At}\sin\omega t, \tag{2}$$

$$\omega = \left(\omega_n^2 - A^2\right)^{1/2}$$

wherein A is the attenuation constant, is the damped frequency of the system, thenatural frequency of the system = ω_n, and $\lambda = 2A/T_d$.

The model parameters λ and T_d are obtained by matching eqn.(1) to the monitored glucose concentration y(t) data (represented by the open circles). The computed values of parameters are: $\lambda = 2.6$ hr-2, $T_d = 1.08$ hr. This computed response is represented in Figure 1 by the bottom curve, fitting the open-circles clinical data.

ParametricIdentification (sample calculation for Normal Test Subject No.5)

$$y\ (1/2) = (G / \omega)e^{-A/2}\sin\omega / 2 = 0.34 \text{ g} / \text{L}$$
$$y\ (1) = (G / \omega)e^{-A}\sin\omega = 0.24 \text{ g} / \text{L}$$
$$y\ (2) = (G / \omega)e^{-2A}\sin 2\omega = -0.09 \text{ g} / \text{L}$$

Using trignometry relations, we get

$$A = 0.8287 \text{ hr}^{-1}$$
$$\lambda = \omega_n^2 = A^2 + \omega^2 = \left(0.82875\right)^2 + \left(2.0146\right)^2 = 4.7455 \text{ hr}^{-2},$$
$$T_d = 2A / \lambda = 0.3492 \text{ hr}$$

Upon substituting the above values of λ and Td, the value of the third parameter,

$$G = 1.2262 \text{ g} \ (1)^{-1} \text{hr}^{-1}$$

For a diabetic subject, the blood-glucose concentration data is depicted by closed circles in Fig 16. For the model to simulate this data, we adopt the solution of model eqn(17), as an over-damped response function:

$$y(t) = (G / \omega) e^{-At}\sinh\omega t \tag{3}$$

The solution (y = (G/ ω) e-Atsinhωt) is made to match the clinical data depicted by closed circles, and the values of λ and T_d are computed to be 0.27 hr-2 & 6.08 hr, respectively. The top curve in **Figure 16** represents the blood-glucose response curve for this potentially diabetic subject. The values of T_d, λ and A for both normal and diabetic patients are indicated in the figure, to provide a measure of difference in the parameter values.

It was found from these calculations that not all of the normal test subjects' clinical data could be simulated as under-damped response. Similarly, not all the diabetic test subjects' clinical data corresponded to over-damped response.

However it was found that the clinical data of these test subjects (both normal and diabetic) could indeed be fitted by means of a critically-damped glucose-response solution of the governing equation.

$$y(t) = G\, t\, e^{-At} \tag{4}$$

for which, $\omega = 0$, $\omega_n{}^2 = A^2 = \lambda$, and $T_d = 2A/\lambda = 2$

Clinically-based Diagnosis:

The blood glucose 'normal' values, used for the clinical studies, were:
Fasting: 70 to 115 mg/dl, At 30th min.: less than 200 mg/dl,
At 1st hour: less than 200 mg/dl, At 2nd hour: less than 140 mg/dl,

Modeling-based Diagnosis:

The test subjects have been classified into four categories:

Normal-test subjects based on under-damped model-response;

Normal test-subjects based on critically-damped model-response, at risk of becoming diabetic;

Diabetic test-subjects based on critically-damped model-response, being border-line diabetic;

Diabetic test-subjects based on over-damped model response;

Non-Dimensional Number for Diagnosis of diabetes:

We decided to develop a unique diabetes index number (DIN) to facilitate differential diagnosis of normal and diabetic states as well as diagnose supposedly normal but high (diabetic) risk patients and diabetic patients in early stages of the disorder [8].

$$\text{DIN} = \frac{y(\max)}{G} \times A \times \frac{T_d}{T(\max)} \tag{5}$$

wherein,

$y(\max)$ = maximum blood glucose value in gram/liter

G = glucose dose administered to the system in gram/liter hour

A = attenuation constant in 1/hour

T_d = derivative-time $(\alpha + \delta)$ in hours

$T(\max)$ = the time at which $y(\max)$ is attained in hour

This non-dimensional number DIN consists of the model parameters (A & T_d) or (A & ω_n) or (λ & T_d). The DIN values for all four categories were computed from equation (5). A distribution plot of the DIN is plotted in fig 17, wherein the DIN is classified into sections with 0.2 increments (for all the four categories of subjects) and the number of subjects which fall into these sections (frequency) is determined.

In the distribution plot (shown in Fig 17), the DIN values 0-0.2 is designated as range 1, the DIN 0.2-0.4 is range 2,   0.4-0.6 is range 3, and so on up to DIN 2.2-2.4, which is range 12.

As can be seen from figure 17, normal (i.e., non-diabetic) subjects with no risk of becoming diabetic, will have DIN value less than 0.4, or be in the 1 - 2 range. Distinctly diabetic subjects will have DIN value greater than 1.2, or be in the 7 - 12 range categories.

Supposedly, clinically-identified normal subjects who have DIN values between 0.4 and 1.0, or are in the 3 - 5 range, are at risk of becoming diabetic.

On the other hand, clinically-identified diabetic subjects with DIN value between 0.4 - 1.2, or in the 3 - 6 range category are border-line diabetics, who can become normal (with diet control and treatment).

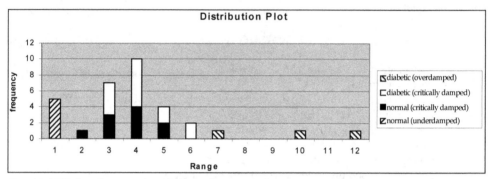

Fig. 17. DIN distribution plot of all the four categories subjects [8]. In the figure, diabetic (critically damped) category of subjects are designated as border-line diabetic; normal (critically damped) category of subjects are designated as normal subjects at risk of becoming diabetic.

7. Cardiology: LV contractility index based on normalized wall-stress

The traditional cardiac contractility index $(dP/dt)_{max}$ requires cardiac catheterization.

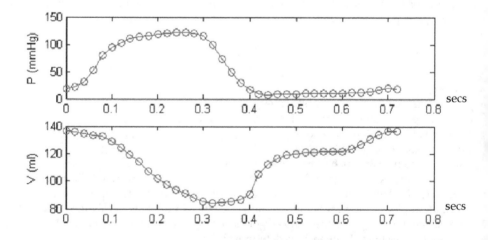

Fig. 18. Sample cyclic variation of LV pressure and volume.

Since LV Pressure is developed by LV wall stress σ_θ(based on sarcomere contraction), we have developed a contractility index based on σ_θ (normalized with respect to LV pressure) [9].

For a thick walled spherical model of LV, the circumferential wall stress:

$$\sigma_\theta(r_i) = P\frac{(r_i^3 / r_e^3 + 0.5)}{(1 - r_i^3 / r_e^3)} \tag{1}$$

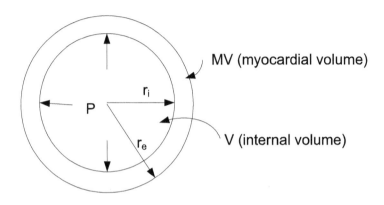

Fig. 19. Thick walled Spherical Model of the LV.

$$\text{The normalised Stress} = \frac{\sigma_\theta(r_i)}{P} = \sigma^* = \frac{3V}{2MV} + \frac{1}{2} \tag{2}$$

We define the contractility index as:

$$\textit{Contractility Index (CONT1)} = \left|\frac{d\sigma^*}{dt}\right|_{max} = \frac{3}{2MV}\left(\frac{dV}{dt}\right)_{max} \tag{3}$$

For the data shown in the figure 18, we have:

$$CONT1 = \frac{3}{200cc}(224cc \cdot s^{-1}) = 3.3s^{-1}$$

Now we formulate a non-dimensional cardiac contractility index,

$$CONT2 = \left|\frac{d\sigma^*}{dt}\right|_{max} \times ejection\ period(= 0.3second) \times 100$$

$$\approx 100$$

Our new contractility indices do not require measurement of LV pressure, and can hence be evaluated noninvasively. In fig 20, we can see how well our contractility index CONT1 $(d\sigma^*/dt)_{max}$ correlates with the traditional contractility index $(dP/dt)_{max}$. This provides a measure of confidence for clinical usage of this index.

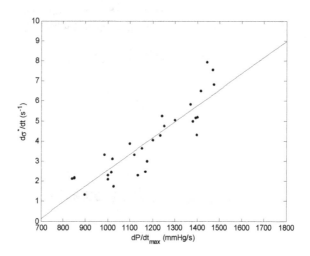

Fig. 20. Correlation of our new contractility index $(d\sigma^*/dt)_{max}$ with the traditional contractility index $(dP/dt)_{max}$

8. Diagnostics: LV contractility index based on LV shape-factor

Cardiologists have observed shape changes taking place in an impaired LV. We have investigated the effect of ventricular shape on contractility and ejection function. In this study, a new LV contractility index is developed in terms of the wall-stress (σ^*, normalized with respect to LV pressure) of a LV ellipsoidal model (Fig. 21) [10, 11].

Using cine-ventriculography data of LV volume (V) and myocardial volume MV, the LV ellipsoidal model (LVEM) major (B) and minor axes (A) are derived for the entire cardiac cycle. Thereafter, our new contractility index $(d\sigma^*/dt)_{max}$ is derived in terms of the LV ellipsoidal shape factor ($s=B/A$).

For the LV model (Fig 21) of a prolate spheroid truncated 50% of the distance from the equator to the base, we first put down the for σ^* ($= \sigma_\theta$ /P), and then determine $d\sigma^*/dt$ [2.10]. Thereby, we obtain the following expression for the contractility index:

$$Contractility\ index - 1\ (CONT1) = \left|(d\sigma^*/dt)\right|_{max}$$

$$= \left| \frac{\dot{V}(2+s)+V\dot{s}}{MV} - \frac{V^2s}{MV(4V+2Vs+MV)^2}\left(\begin{array}{c} s\dot{V}(8+4MV/V+(8+2MV/V)s+2s^2) \\ +V\dot{s}(16+4MV/V+(16+3MV/V)s+4s^2) \end{array} \right) \right|_{max} = F(s,\dot{s},V,\dot{V},MV)$$

where s is B/A, $\dot{s}$ is first-time derivative of s; V and MV are LV volume and myocardial wall volume, $\tilde{V}$ is the first-time derivative of V.

This index is analogous to the traditional employed index $(dP/dt)_{max}$, but does not involve determination of the intra-LV pressure by catheterization. For patient A (with myocardial infarct and double vessel disease) and B (with double vessel disease and hypertension), the values of CONT1 are obtained to be 3.84 and 6.90 s^{-1}, whereas the corresponding values of $(dP/dt)_{max}$ are obtained to be 985 and 1475 mmHg/s.

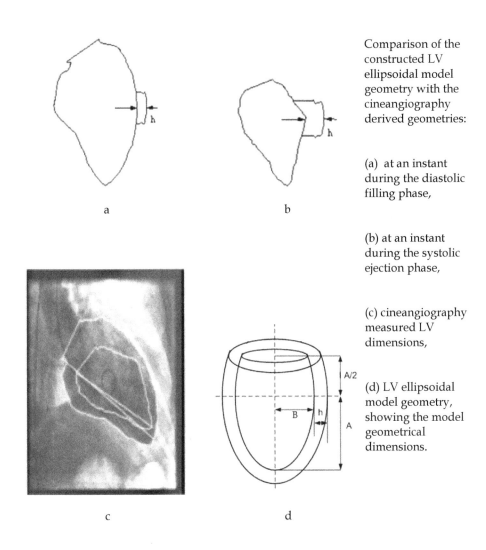

Comparison of the constructed LV ellipsoidal model geometry with the cineangiography derived geometries:

(a) at an instant during the diastolic filling phase,

(b) at an instant during the systolic ejection phase,

(c) cineangiography measured LV dimensions,

(d) LV ellipsoidal model geometry, showing the model geometrical dimensions.

Fig. 21. Cineangiography imaged LV geometry and the corresponding constructed LV ellipsoidal model: (a) at an instant during the diastolic filling stage, (b) at an instant during the systolic ejection stage, (c) measured LV dimensions, (d) LV ellipsoidal model, depicting its geometrical parameters.

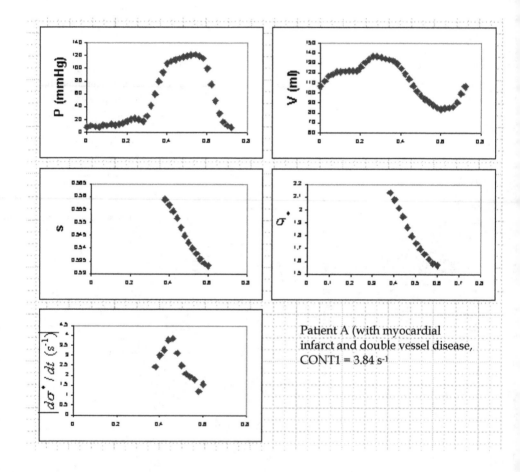

Patient A (with myocardial infarct and double vessel disease, CONT1 = 3.84 s⁻¹

Fig. 22. Patient A: Cyclic variation of LV pressure-volume data, LV model shape factor s, computed σ^*, computed index $(d\sigma^*/dt)_{max}$.

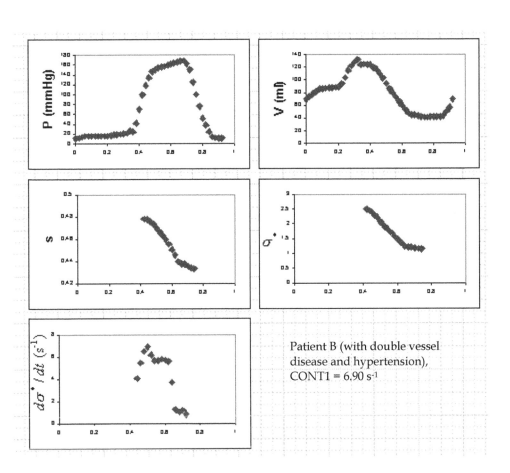

Patient B (with double vessel disease and hypertension), CONT1 = 6.90 s⁻¹

Fig. 23. Patient B: Cyclic variation of LV pressure-volume data, LV model shape factor s, computed σ^*, computed index $(d\sigma^*/dt)_{max}$.

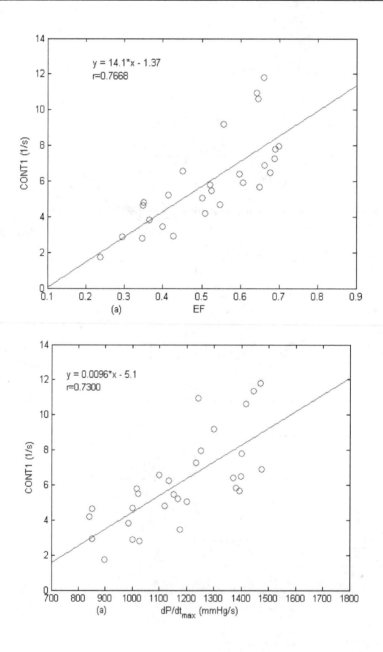

Fig. 24. Correlationships between (i) CONT1 and EF, and (ii) CONT1 and *(dP/dt)max*.

For our patient group, we have computed and plotted CONT1 vs EF and $(dP/dt)_{max}$ in Fig. 24. We can note the good level of correlation with the traditional contractility index of $(dP/dt)_{max}$. Additionally, our new index can be determined noninvasively, and hence be more conducive for clinical use.

From these results, we can also infer that a non-optimal less-ellipsoidal shape (or a more spherical shape, having a greater value of $S = B/A$) is associated with decreased contractility (and poor systolic function) of the LV, associated with a failing heart. This has an important bearing on a quick assessment of a failing heart based on the values of S and $(d\sigma^*/dt)_{max}$

9. ICU Evalution: Indicator for lung-status in mechanically ventilated copd patients (using lung ventilation modelling and assessment)

In chronic-obstructive-pulmonary-disease (COPD), elevated airway resistance and decreased lung compliance (i.e. stiffer lung) make breathing difficult. After these patients are mechanically ventilated, there is a need for accurate predictive indicators of lung-status improvement, for ventilator discontinuation through stepwise reduction in mechanical support, as and when patients are increasingly able to support their own breathing, followed by trials of unassisted breathing preceding extubation, and ending with extubation.

So, in this section, we have provided a biomedical engineering analysis of the lung ventilator volume response to mechanical ventilation of COPD patients, and developed an index to assess the lung status as well as the basis of weaning the patient from ventilator support.

Figure 25 depicts the model for the lung volume (V) response to the net driving pressure $P_N = P_L - P_e$ (end-expiratory pressure), in which (i) the *driving pressure* $P_L = P_m$ (*pressure at the mouth*) minus P_p (*the pleural pressure*), and (ii) P_p is determined by intubating the patient, and assuming that the pressure in the relaxed esophageal tube equals the pressure in the pleural space surrounding it.

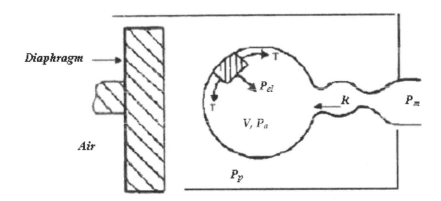

Fig. 25. Model of the Lung, depicting P_m, P_p, P_{el} (lung elastic pressure recoil) = P_a (alveolar pressure) – P_p (pleural pressure) = 2T/(radius of alveolar chamber), and R (resistance to airflow rate) = $(P_m - P_a) / (dV/dt)$.

The equation representing the lung model response to the net driving pressure in terms of the model parameters lung compliance (C) and airflow-resistance (R), is given by:

$$R \overset{o}{V} + \frac{V}{C} = P_L(t) - P_e = P_N(t) \tag{1}$$

wherein:

1. the driving pressure = P_L ; Pe = the end-expiratory pressure; net pressure $P_N = P_L - P_e$
2. the parameters of the governing equation (1) are lung compliance (C) and airflow-
3. resistance(R), with both R & C being instantaneous values
4. V= V(t) - Ve (wherein Ve is the end expiratory lung volume)

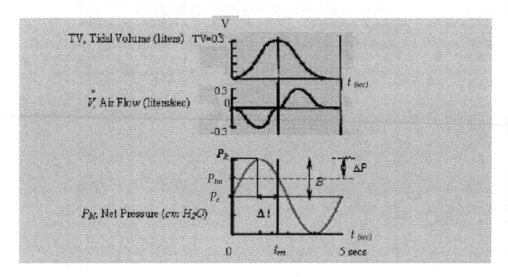

Fig. 26. Lung ventilatory model data shows air-flow ($\overset{o}{V}$) and volume (V) and net pressure (P_N). Pause pressure (P_{tm}) occurs at t_m, at which the volume is maximum (TV = tidal volume). Δt is the phase difference between the time of maximum volume and peak pressure (P_k). It also the time lag between the peak and pause pressures. B is the amplitude of the net pressure waveform P_N applied by the ventilator. This P_N oscillates about P_e with amplitude of B. The difference between peak pressure P_k and pause pressure P_{tm} is Δp [12].

We measure Peak pressure (P_k), Pause pressure (P_{tm}), t_m & ω (or $t_m \omega$). Then,

$$P_k = P_L (t = \pi / 2\omega) = P_e + B$$
$$P_{tm} = P_L [t = t_m = (\pi - \theta) / \omega] = P_e + B \sin \omega \, t_m = P_e + B \sin [\omega \, (\pi-\theta)/\omega]$$
$$= P_e + B\sin(\pi - \theta) = P_e + B\sin\theta$$
$$\therefore B = (P_k - P_{tm})/(1 - \sin\theta) = \Delta P/(1-\sin\theta)$$

B is the amplitude of the net pressure wave form applied by the ventilator. Let C_a be the average dynamic lung compliance, R_a the average dynamic resistance to airflow, the driving pressure $P_L = P_e + B \sin(\omega t)$, and the net pressure $P_N = B \sin(\omega t)$. The governing equation (1) then becomes [12]:

$$R_a \overset{o}{V} + \frac{V}{C_a} = P_N = B\sin(\omega t) \qquad (2)$$

The volume response to P_N, the solution to eqn (2), is given by:

$$V(t) = \frac{BC_a \{\sin(\omega t) - \omega k_a \cos(\omega t)\}}{1 + \omega^2 k_a^2} + He^{-t/k_a} \qquad (3)$$

wherein:
   i.   ka (=RaCa)is the average time constant,
   ii. the integration constant H is determined from the initial conditions,
   iii. the model parameters are Ca and ka(i.e. Ca and Ra), and
   iv. ω is the frequency of the oscillating pressure profile applied by the ventilator
An essential condition is that the flow-rate is zero at the beginning of inspiration and end of expiration. Hence, applying this initial condition of $dV/dt = 0$ at $t=0$ to our differential eqn (3), the constant H is obtained as:

$$H = \frac{BC_a \omega k_a}{1 + \omega^2 k_a^2} \qquad (4)$$

Then from eqn (3) & (4), we obtain:

$$V(t) = \frac{BC_a \{\sin(\omega t) - \omega k_a \cos(\omega t)\}}{1 + \omega^2 k_a^2} + \frac{BC_a \omega k_a}{1 + \omega^2 k_a^2} e^{-t/k_a} \qquad (5)$$

Evaluating parameters R_a & C_a [12,2]:

For evaluating the parameter $k_a(R_a C_a)$, we will determine the time at which V (t) is maximum and equal to the tidal volume (TV), Hence putting $dV/dt = 0$ in eqn (5), we obtain:

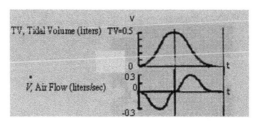

$$\cos(\omega t) + \omega k_a \sin(\omega t) = e^{\left(\frac{-t}{k_a}\right)}, \text{ at } t = tm \qquad (6)$$

From equation (6), we obtain (by neglecting e^{-t/k_a}), the following expression for k_a:

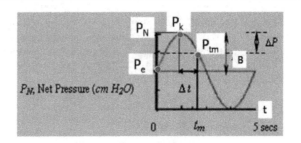

$$k_a = -1/\omega \tan(\omega t_m) \tag{7a}$$

$$or, \quad \tan^{-1}(1/\omega k_a) = \pi - (\omega t_m) = \theta \tag{7b}$$

From eqn (5 & 6):

$$V(t = t_m) = TV = \frac{BC_a \{\sin(\omega t_m) - \omega k_a \cos(\omega t_m)\}}{1 + \omega^2 k_a^2} + \frac{BC_a \omega k_a}{1 + \omega^2 k_a^2} e^{-tm/ka} \tag{8}$$

At $t = t_m$, the second term,

$$H = \frac{BC_a \omega k_a}{1 + \omega^2 k_a^2} e^{-tm/ka} \approx 0 \tag{9}$$

Hence, eqn (8) becomes:

$$V(t = t_m) = TV = \frac{BC_a \{\sin(\omega t_m) - \omega k_a \cos(\omega t_m)\}}{1 + \omega^2 k_a^2} \tag{10}$$

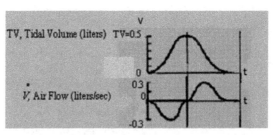

In eqn (10), if we put:

$$N = \sin(\omega t_m) - \omega k_a \cos(\omega t_m) \tag{11}$$

Then, based on equations (6 and 7), we get:

$$N = \frac{1}{\sqrt{1 + \omega^2 k_a^2}} + \frac{\omega^2 k_a^2}{\sqrt{1 + \omega^2 k_a^2}} = \sqrt{1 + \omega^2 k_a^2} \tag{12}$$

Now then, based on equation (12), equation (10) becomes:

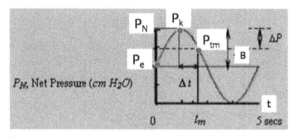

Fig. 26. (reproduced)

$$V(t=t_m) = TV = \frac{BC_a}{\sqrt{1+\omega^2 k_a^2}}$$ (13)

Determining Lung-Compliance (Ca) and Air-Flow Resistance (Ra):

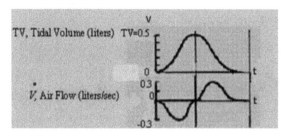

From equations (13 and 7), we get:

$$C_a = \frac{TV\sqrt{1+\omega^2 k_a^2}}{B} = \frac{TV}{B\sin\theta} = \frac{TV(1-\sin\theta)}{\Delta P \sin\theta}$$ (14)

Hence, from eqns (7 & 13), the average value of airflow-resistance (R_a) is:

$$R_a = k_a / C_a = \frac{\Delta P \sin\theta(1/\omega\tan\theta)}{TV(1-\sin\theta)} = \frac{\Delta P \cos\theta}{TV\omega(1-\sin\theta)}$$ (15)

For our patients, the computed ranges of the parameters are:

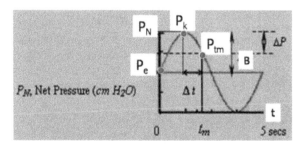

$$C_a = 0.020 - 0.080 \; L \, / \, cmH_2O \qquad (16)$$

$$R_a = 9 - 43 \; cmH_2O \cdot s \, / \, L$$

Now, that we have determined the expressions for R_a and C_a, the next step is to develop an integrated index incorporating these parameters.

Formulating a Lung Ventilatory Index (LVI) incorporating Ra and Ca: We now formulate a Lung Ventilatory Index (LVI), incorporating Ra and Ca, as:

$$LVI = \frac{R_a(RF)P_k}{C_a(TV)} \qquad (17)$$

Let us now obtain order-of-magnitude values of LVI, for a mechanically ventilated COPD patient in acute respiratory failure:

$$LVI(Intubated \; COPD) = \frac{[15 cmH_2Os \, / \, L][0.33_s{}^{-1}][20 cmH_2O]}{[0.035L \, / \, cmH_2O][0.5L]}$$

$$= 5654 \; (cmH_2O \, / \, L)^3$$

wherein

$$R_a = 15 \, cmH_2Os \, / \, L \qquad C_a = 0.035L \, / \, cmH_2O \qquad RF = 0.33_s{}^{-1}$$

$$TV = 0.5L \qquad\qquad P_k = 20 cmH_2O$$

Now, let us obtain an order-of-magnitude of LVI (by using representative computed values of R_a, C_a, RF, TV, and P_k) as above for a COPD patient with improving lung-status just before successful discontinuation.

$$LVI(Outpatient \; COPD) = \frac{[10 cmH_2Os \, / \, L][0.33_s{}^{-1}][12 cmH_2O]}{[0.05L \, / \, cmH_2O][0.35L]}$$

$$= 2263 (cmH_2O \, / \, L)^3$$

wherein

$$R_a = 10 \, cmH_2Os \, / \, L \qquad C_a = 0.050L \, / \, cmH_2O \qquad RF = 0.33_s{}^{-1}$$

$$TV = 0.35L \qquad\qquad P_k = 12 cmH_2O$$

Hence, for LVI to reflect lung-status improvement in a mechanically ventilated COPD patient in acute respiratory failure, it has to decrease to the range of LVI for an outpatient COPD patient at the time of discontinuation.

In fig. 27, it is shown that for the 6 successfully-discontinued cases, the LVI was (2900) ± (567) $(cmH_2O/L)^3$; for the 7 failed-discontinuation cases the LVI was (11400) ± (1433) $(cmH_2O/L)^3$. It can be also observed that LVI enables clear separation between failed and successful discontinuation, which again points to the efficacy of LVI.

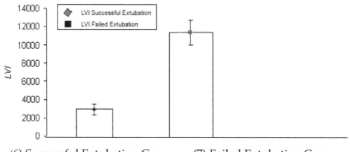

(6) Successful Extubation Case   (7) Failed Extubation Case

Fig. 27. Distribution of LVI at discontinuation for patients with failed and successful discontinuation. For the 6 successfully-discontinued cases, the LVI was (2900) ± (567) $(cmH_2O/L)^3$; for the 7 failed-discontinuation cases the LVI was (11400) ± (1433) $(cmH_2O/L)^3$. It is observed that LVI enables clear separation between failed and successful discontinuation [12].

10. Monitoring: Noninvasive determination of aortic pressure, aortic modulus (stiffness) and peripheral resistance)

The aortic blood pressure waveform contains a lot of information on how the LV contraction couples with the aortic compliance and peripheral resistance. Since accurate measurement of aortic blood pressure waveform requires catheterization of the aorta, we have developed a noninvasive method to determine the aortic pressure profile along with the aortic volume-elasticity and peripheral resistance. Fig 28 displays such a constructed aortic pressure profile.

The input to the model consists of auscultatory cuff diastolic and systolic pressures, along with the MRI (or echocardiographically) measured ejection volume-time profile (or volume input into the aorta). The governing differential equation for pressure response to LV outflow rate $I(t)$ into the aorta is given by [13]:

$$\frac{dP}{dt} + \lambda P = m_a \left[I(t) + T_a \frac{dI}{dt} \right]$$

(1)

where (i) m_a is the aortic volume elasticity (dP/dV), (ii) R_p is the resistance to aortic flow $(=P/Q)$, (iii) $\lambda = m_a/R_p$, (iv) $I(t)$ is the monitored inflow rate into the aorta, and (v) T_a is the flow-acceleration period.

This governing equation is solved for measured $I(t)$ and dI/dt during the systolic phase from time T_1 to T_3 (Fig. 28). For the diastolic-phase solution from time T_3 to T_4, the right-hand side is zero. The solutions for diastolic and systolic phases are given below by equations (2) and (3) respectively.

Solutionequations:

• During diastolic phase,

$$P_d(t) = P_1 e^{\lambda(T-t)}$$

(2)

At $T = T_4$, $P_d(T)$=auscultatory $P_{ad} = P_1$

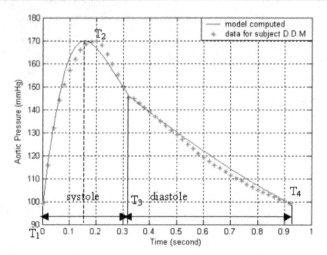

Fig. 28. Model computed cyclic Aortic pressure profile, compared with measured pressure values. The systolic phase from $T1$ to $T3$ is when blood is ejected into the aorta. The diastolic phase is from $T3$ to $T4$ [13].

- During systolic phase,

$$P_s(t) = (P_1 - A_1)e^{-\lambda t} + e^{-bt}\left[A_1 \cos(\omega t) + A_2 \sin(\omega t)\right] \tag{3}$$

where

$$A_1 = \frac{m_a a\omega(T_a\lambda - 1)}{(b-\lambda)^2 + \omega^2}$$

$$A_2 = \frac{m_a a[(\omega^2 + b^2)T_a + \lambda - b - T_a b\lambda]}{(b-\lambda)^2 + \omega^2}$$

Also, as noted in Fig 28, the boundary values are:
At t $=T_2$, dP_s/dt =0; at t $=T_2$, $P_s(T_2)$ = auscultatory P_{as}= P_2 ; at t = T_3, $P_s(t=T_3)$ = $P_d(t=T_3)$
Hence, based on these boundary values, the following equations are to be solved.

$$\frac{dP_s(t)}{dt}(t=T_2) = -C_1\lambda e^{-\lambda T_2} + e^{-bT_2}\left[A_3 \cos(\omega T_2) - A_4 \sin(\omega T_2)\right] = 0 \tag{4}$$

- At $t = T_2$, $P_s = P_2$. Hence, from equations (3 & 4), we get:

$$P_s(t=T_2) = P_2 = C_1 e^{-\lambda T_2} + e^{-bT_2}\left[A_1 \cos(\omega T_2) + A_2 \sin(\omega T_2)\right] \tag{5}$$

- At $t = T_3$, $P_s(t=T_3)$ = $P_d(t=T_3)$. Hence, from equations (2 and 3), we obtain:

$$P_1 e^{\lambda(T-T_3)} = (P_1 - A_1)e^{-\lambda T_3} + e^{-bT_3}\left[A_1 \cos(\omega T_3) + A_2 \sin(\omega T_3)\right] \tag{6}$$

We now solve equations (4, 5 & 6), to determine the unknown parameters m_a, R_p (and T_2).

We have determined the expressions for the aortic pressure during the systolic and diastolic phases, by solving the governing equation (1), for the monitored LV outflow rate (or input into the aorta) $I(t)$, using (i) the monitored auscultatory diastolic pressures (P_{ad}), to serve as the boundary condition at the beginning of the systolic-phase solution (at time T_1) and at the end of the diastolic-phase solution (at time T_4), (ii) the monitored auscultatory systolic pressure (P_{as} =P_2) to represent the maximum value of the systolic-phase solution.

Because the pressure solution of equation (1) is a function of m_a and R_p, we first determine the values of these parameters by making the solution satisfy the above-mentioned boundary conditions, which in turn yields the pressure profile given by $P_d(t)$ and $P_s(t)$. This non-invasively determinable aortic pressure can provide hitherto unavailable information on vascular compliance and resistance status, as well as on the capacity of the LV to respond to it.

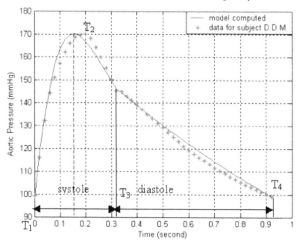

Fig. 29. Plot of aortic pressure (of patient B) during one cardiac cycle, t_e=0.32s. Herein, T_1-T_3 represents the systolic phase, and T_3-T_4(=0.92s) represents the diastolic phase. At T_2, the aortic systolic pressure profile has its maximal value (=P_s). The scatter points are the data measured from catheterization. The solid line is the model-computed profile; RMS=2.41 mmHg. Note the excellent correlation between the model-derived aortic pressure profile and the catheter-obtained aortic pressure profile.

In **Fig. 29**, the model-computed aortic pressure profile patient B (with double vessel disease and hypertension) is shown, along with the actual catheter pressure data. We can note how well the model-computed result matches the actual catheterization data, with RMS 2.41 mmHg. The aortic stiffness (m_a) and peripheral resistance (R_p) are obtained to be 1.03 $mmHg/ml$ and 1.59 $mmHg \cdot s/ml$, respectively.

Let us consider yet another benefit of this analysis. We have determined aortic pressure profile, aortic stiffness (m_a) or aortic elastance (E_{ao}) and peripheral resistance. From the instantaneous aortic pressure and aortic inflow rate, we can determine the instantaneous left ventricular (LV) systolic pressure, in terms of the instantaneous dimensions of the LV outflow tract. We hence determine the LV systolic pressure profile, from which we can evaluate the traditional contractility index (dP/dt max) as well as the LV systolic elastance (Elv).

We can then determine the ratio of Elv/Eao, to represent the LV-Aorta Matching Index ($VAMI$). In ischemic cardiomyopathy patients, this $VAMI$ value is depressed. However, following surgical vascular restoration, this index value is partially restored.

11. Coronary Bypass surgery: Candidacy

- Interventional candidacy based on computed intra-LV flow-velocity and pressure-gradient distributions

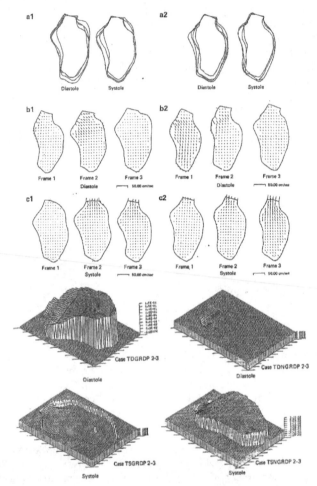

Fig. 30. Construction of Intra-LV Blood flow velocity and pressure-gradient distributions for a patient with myocardial infarct: (a) Superimposed sequential diastolic and systolic endocardial frames (whose aortic valves centers and the long axis are matched), before (1) and after (2) administration of nitroglycerin; (b) Instantaneous intra-LV distributions of velocity during diastole, before (1) and after (2) administration of nitroglycerin; (c) Instantaneous intra-LV distributions of velocity during ejection phase, before (1) and after (2) administration of nitroglycerin, (d) Instantaneous intra-LV distributions of pressure-differentials during diastole, before (1) and after (2) administration of nitroglycerin; (e) Instantaneous intra-LV distributions of pressure-differential during ejection phase, before (1) and after (2) administration of nitroglycerin. (Adapted from reference 14: Subbaraj K, Ghista DN, Fallen EL, *J Biomed Eng* 1987, 9:206-215.).

A left ventricle with ischemic and infarcted myocardial segments will have lowered ejection fraction and cardiac output, because it will not be able to generate adequate myocardial contraction to raise the intra-LV pressure above aortic pressure for a ling enough duration to generate adequate stroke volume. These patients need coronary bypass surgery, and a pre-surgical assessment of their candidacy for it, on how much they can from it. For this purpose, we need to determine the intra-LV blood flow velocity and pressure-gradient profiles before and after the administration of nitroglycerin.

So, we carry out a CFD analysis of intra-LV blood flow. The data required for the CFD analysis consists of: LV 2-D long-axis frames during LV diastolic and systolic phases; LV pressure vs. time associated with these LV frames; Computation of LV instantaneous wall velocities as well as instantaneous velocity of blood entering the LV during the filling phase and leaving the LV during the ejection phase.

From this CFD, we have determined the instantaneous distributions of intra-LV blood-flow velocity and differential-pressure during filling and ejection phases, to intrinsically characterize LV resistance-to-filling (LV-RFT) and contractility (LV-CONT) respectively. The results are summarized in the above Fig. 30.

By comparing intra-LV pressure-gradients before and after administration of nitroglycerin (a myocardial perfusing agent, and hence a quasi-simulator of coronary bypass surgery), we can infer how the myocardium is going to respond and how these LV functional indices will improve after coronary by bypass surgery.

12. Theory of hospital administration: Formulation of hospital units' performance index and cost-effective index

A Hospital has clinical services departments, medical supply and hospital-services departments, financial-management and administrative departments. Each of these five sets of departments has to function in a cost-effective fashion.

Let us, for example, consider the Intensive-Care Unit (ICU) department. The human resource to an ICU dept consists of physicians and nurses. Using activity-based costing, we can determine the human-resource strength, based on an assumed reasonable probability-of-occurrence of (for instance) two patients simultaneously (instead of just one patient) having life-threatening episodes.

Performance Index: We can formulate the ICU performance-indicator in terms of the amounts by which the physiological health-index (PHI or NDPI) values of patients were (i) enhanced in the ICU for those patients discharged into the ward from the ICU, and (ii) diminished in the ICU in the case of patients who died in the ICU.

Let us say that patients are admitted to the CCU if their Physiological-health-index (PHI) value falls below 50%. So if the PHI of a patient improves from 30 to 50, then the **Patient-HealthImprovement Index (PHII)** for that patient is given by [15],

$$\text{PHII} = \left(\frac{50 - 30}{30} \right) 100 = 67 \ (or \ 67\%) \tag{1}$$

Thus the **patient health-improvement index (PHII)** value is higher if a more seriously-ill patient is discharged from the ICU, and lower if a not-so-seriously ill patient is discharged, i.e., if

$$\text{PHII} = \left(\frac{50 - 40}{40} \right) 100 = 25 \ (or \ 25\%) \tag{2}$$

We can then formulate the **Performance-index (PFI)** for an ICU as follows:

$$ICU \text{ Performance index } (PFI) = \frac{\Sigma \; PHII \text{ of the patients}}{\text{\# of those patients treated during a life time period}} \qquad (3)$$

Hence, the higher the value of ICU performance index, the better is the performance of the ICU. If now a patient dies, as a result of the PHII becoming negative (i.e slipping from (say) 30 to 10), then

$$PHII = 100 \left(\frac{10-30}{30} \right) = -67 \qquad (4)$$

As a result, ΣPHII (in equation 3) will decrease, and the overall value of ICU performance index (namely PFI, as calculated by means of eq. 3), will fall.

Cost-Effective Index: Now, let us consider that (i) we have one physician and five nurses for a 10 bed CCU, based on the probability-of-occurrence of two patients having life-threatening events being say 0.2 (or 20%), and that (ii) for this human resource/staffing, the ICU performance index value (PFI) is (say) 40.

If we increase the staffing, the ICU performance index value could go upto 50 or so, at the expense of more salary cost. So now we can come up with another indicator namely, **Cost-effectiveness index (CEI)**

$$CEI = \frac{\text{Performance index}}{\text{Total salary index (in salary - units)}}$$

$$= \frac{\text{Performance index}}{\text{Resource index (in terms of salary - units)}} \qquad (5)$$

wherein, say, a salary of 1000=0.1 unit, 10,000=1 unit, 20,000=2 units, and so on.
So if an ICU has one physician with a monthly salary of 20,000 (i.e. 2 salary-units) and four nurses each with a total monthly salary of 5,000 (i.e. total of 2.0 salary-units), then

$$\mathbf{CEI(ICU)} = \frac{\text{Performance index (of 40)}}{\text{Salary-units Index or Resource Index,} R_i \, [=(2+2.0)]}$$

$$= \frac{40}{4.0} = 10 \qquad (6)$$

Now, let us assume that we raise the **PFI** (ICU) to (say) 60 by augmenting the nursing staff, so as to have six nurses (R_i = 3 units) and 1.5 full-time equivalent physician-on-duty (R_i = 3 units), then

$$CEI \; (ICU) = \frac{PFI}{R_i} = \frac{50}{(3+3)} = 8.3 \qquad (7)$$

So while the PFI of ICU has gone up from 40 to 50, the CEI of ICU has gone down from 10 to 8.3.

Strategy of Operation: Our strategy would be to operate this Performance-Resource system, in such a way that we can determine the resource index R_i for which we can obtain acceptable values of both PFI and CEI. In a way, figure (31) could represent this balance between PFI & CEI, in order to determine optimal Resource Index or resource [15].

With reference to this **figure 31**, if we have a resource value of $R_i = R_1$, then the corresponding PFI (=PFI_1) will be unacceptable, as being too low; hence, we will want to increase the value of R_i. If we have a resource value of $R_i = R_2$, then our corresponding CEI (= CEI_2) will also be unacceptable, for being too low; hence we will want to decrease R_i. In doing so, we can arrive at the optimal value R_{io}, for which both CEI and PFI are acceptable. This procedure, for converging to $R_i = R_{io}$, can be formulated computationally.

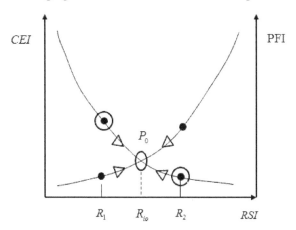

Fig. 31. Optimising the value of R_i so to obtain acceptable values of CEI and PFI [15].

Now then, let us formulate how a hospital budget can be optimally distributed. Let us say that a Hospital has 'n' number of departments and a prescribed budget (or budget index, BGI). We would want to distribute the budget among the departments, such that none of the 'n' departments has a PFI below the acceptable value of PFI_a and a CEI below the acceptable value of CEI_a.

So the Operational research problem is to be formulated as follows:

How to distribute or divide the given Budget (or Budget Index Value) into R_i (i= 1,……..,n), such that $PFI_i \geq PFI_a$, for all i; and $CEI_i \geq CEI_a$, for all i This then is the prime task of a Hospital administrator!

Summarizing:

The ICU Operational cost (OPC) is the cost of operating the ICU over this one month period. The ratio of this ICU PFI and the ICU OPC is the Cost-effectiveness index (CEI) of ICU.
The Management strategy is to maintain certain acceptable values of both PFI and CEI for all hospital departments, by judicious allocation of staff to the departments.
This enables the determination of optimal Resource index (RSI) and hospital budget (HOB) to maintain a balance between PFI and CEI for all the hospital departments. This can constitute the basis of Hospital Management.

13. *How to proceed*: For biomedical engineering to become a professional field

So in this chapter, we have talked about:
1. *Anatomy:* Spine as an optimal structure,
2. *Physiology:* Mechanism of LV pressure generation

3. Non-dimensional Physiological Index
4. *Medical Test:* Cardiac Fitness based on Treadmill Heart Rate Variation
5. *Medical Physiology:* A Non-dimensional Diabetes Index with respect to Oral-Glucose-tolerance Testing
6. *Cardiology:* LV Contractility based on its normalized wall stress
7. *Diagnostics:* LV Contractility based on its shape factor
8. *ICU Evaluation:* Indicator for lung-status in Mechanically ventilated Copd patients, using lung ventilation modeling and assessment
9. *Monitoring:* Noninvasive determination of Aortic pressure, aortic stiffness and peripheral resistance
10. *Coronary Bypass Surgery:* candidacy
11. Hospital Operations Management

We have shown how biomedical engineering (BME) can open up a new approach to the study of Anatomy, in terms of how anatomical structures are intrinsically designed as optimal structures for their function.

We have also seen how BME can be applied to the study of Physiology, by modelling of physiological systems (as bme models), which can enable us to assess their performance for clinical usage by means of our novel NDPIs comprised of the physiological systems' model parameters. The NDPIs can considerably facilitate medical assessment, and lead us to what can be effectively termed as Higher-order Translational Medicine (HOTM), entailing the incorporation of physical and engineering sciences into medical sciences and clinical sciences for more reliable and effective patient care.

Even Medical tests can be adroitly modelled as BME systems. These systems are formulated in the form of mathematical models, whose solutions are simulated to the medical tests data, to evaluate the model parameters. The model parameters can be combined into NDPIs, by means of which the test data can be analysed for making more reliable medical assessment and decisions.

Finally, we have seen how we can bring to bear the enormous scope of Industrial Engineering (and constrained optimisation theory) to hospital operational management, so as to develop (i) cost-effective performance indices of hospital departments, and (ii) more knowledgeable framework for budget development and allocation to the various department.

Putting all of this together is what will justify (i) the incorporation of BME into the MD curriculum, and (ii) its constituting an indispensable patient-care oriented department in tertiary- care hospitals.

This verily constitutes the biomedical engineering professional trail, from anatomy and physiology to medicine and into hospital administration. This is what the professional role of biomedical engineering needs to be, to promote a higher order of translational medicine and patient care.

14. References

[1] Nondimensional Physiological Indices for Medical assessment, by Dhanjoo N. Ghista, in *J of Mechanics in Medicine and Biology,* Vol 9, No 4, 2009.

[2] *Applied Biomedical Engineering Mechanics*, by Dhanjoo N. Ghista, CRC press (Taylor & Francis Group) Baton Rouge Florida 334872-2742, ISNBN 978-0-8247-5831-8,2008

[3] Human Lumbar Vertebral Body as an Intrinsic Functionally-optimal Structure, by D.N. Ghista, S.C. Fan, K.Ramakrishna, I.Sridhar, in *International Journal of Design and Nature*, 2006, 1(1): 34-47.

[4] The Optimal Structural design of the human Spinal Intervertebral disc", by D.N.Ghista, S.C.Fan, I.Sridhar, K.Ramakrishna, in *International Journal of Design and Nature*, 2006, 1(2).

[5] Mechanism of Left Ventricular Pressure increase during Isovolumic contraction, and determination of its Equivalent myocardial fibers orientation, by Ghista, DN, Liu Li, Chua LP, Zhong L, Tan RS, Tan YS; in *J Mech Med Biol*, 2009, 9 (2), 177–198

[6] Cardiac Fitness mathematical Model of Heart-rate response to V02 during and after Stress-Testing", Lim GeokHian, Dhanjoo N. Ghista, Koo TseYoong, John Tan Cher Chat, Philip EngTiew& Loo Chian Min; *International Journal of Computer Application in Technology(Biomedical Engineering & Computing Special Issue)*, Vol 21, No 1/2, 2004.

[7] Glucose Tolerance Test Modeling & Patient-Simulation for Diagnosis, by Sarma Dittakavi & Dhanjoo N. Ghista, *Journal of Mechanics in Medicine & Biology*, Vol. 1, No.2, Oct.2001.

[8] Clinical Simulation of OGTT Glucose Response Model for Diagnosis of Diabetic Patient, by Dhanjoo N.Ghista, Patrick S.K. Chua, Andy UtamaAulia, Peter L.P. Yeo, in *Journal ofMechanics in Medicine & Biology* 2005 Vol 5, No. 1.

[9] Validation of a novel noninvasive characterization of cardiac index of left ventricle contractility in patients, by Zhong L, Tan RS, Ghista DN, Ng E. Y-K, Chua LP, Kassab GS, *Am J Physiol Heart CircPhysiol*2007, 292:H2764-2772.

[10] LV shape-based contractility indices, by Liang Zhong, Dhanjoo N. Ghista, Eddie Y-K. Ng, Lim SooTeik and Chua Siang Jin, CN Lee, in *Journal of Biomechanics*, 2006, 39: 2397-2409.

[11] Measures and Indices for Intrinsic Characterization of Cardiac Dysfunction during Filling and Systolic Ejection, by Liang Zhong, Dhanjoo N. Ghista, Eddie Y. Ng, Lim SooTeik, and Chua Siang Jin, in *Journal of Mechanics in Medicine and Biology*,Vol 5, No. 2, 2005.

[12] Indicator for Lung-status in a mechanically Ventilated (COPD) Patient using Lung Ventilation Modeling and Assessment, by D.N.Ghista, R. Pasam, S.B. Vasudev, P.Bandi, and R.V. kumar in *Human Respiration Anatomy and Physiology, Mathematical Modellings and Applications,* ed by Vladimir Kulish, WIT Press, 2006.

[13] Determination of Aortic Pressure-time Profile, Along with Aortic Stiffness and Peripheral Resistance, by Liang Zhong, Dhanjoo N. Ghista, Eddie Y-K. Ng, Lim SooTeik and Chua Siang Jin, in *Journal of Mechanics in Medicine & Biology* 2004, 4(4):499-509.

[14] Intrinsic Indices of the Left Ventricle as a Blood Pump in Normal and Infarcted Left Ventricle, by K. Subbaraj, D.N. Ghista, and E. L. Fallen, in *J of Biomedical Engineering*, Vol 9, July issue, 1987

[15] Physiological Systems' Numbers in Medical Diagnosis and Hosipital Cost-effective Operation, by Dhanjoo N. Ghista, in *Journal of Mechanics in Medicine & Biology* 2005, vol 4, No.4.

Chemical Carcinogenesis: Risk Factors, Early Detection and Biomedical Engineering

John I. Anetor[1], Gloria O. Anetor[2], Segun Adeola[1] and Ijeoma Esiaba[1]
[1]Department of Chemical Pathology, College of Medicine, University of Ibadan,
[2]Department of Human Kinetics and Health Education, Faculty of Education,
University of Ibadan,
Nigeria

1. Introduction

Cancer is now recognized in both humans and in other multicellular animals as arising from a number of different causes, including specialized viruses, radiation, chemicals, certain highly irritative parasites (inflammation) and a number of other factors, such as specific genetic defects present in individual humans and possibly in every member of a colony of specially bred animal models. Cancer from non genetic causes largely from environmental factors, of which chemicals have a disproportionate share, is believed to contribute nearly 70% of all cancer cases. Chemical carcinogenesis originally derives from experimental induction of malignant skin tumor in mice with chemicals. Early studies indicated some agents such as polycyclic aromatic hydrocarbons (PAH) could cause cancer of the skin if they were painted on to mice in high doses. These early studies also showed that the induction of cancer was dose dependent; in low dosage they would not cause cancer but would render the skin susceptible to developing cancer on exposure to another agent, which, on its own would not induce cancer.

Thus at the dawn of the 20th century, it was recognized that chemicals cause cancer; though individual cancer causing molecules had not yet been identified, nor their cellular targets clearly known. It was however clearly understood that carcinogenesis, at the cellular level, was predominantly an irreversible process. Knowledge of the mechanisms by which chemicals cause cancer and the molecular changes that characterize tumor progression was lacking. The origin of the understanding that cancer had a cause was first pointed out by the Italian investigator, Ramazini in 1700. Seven and a half decades later, the British Surgeon, Percival Pott made the connection between exposure to soot, rich in hydrocarbons and scrotal cancer (Pott, 1775). It is now known that at the most fundamental level, cancer is caused by abnormal gene expression. This abnormal gene expression occurs through a number of mechanisms, including direct damage to the DNA, and inappropriate transcription and translation of cellular genes. The contribution of chemicals to the carcinogenic process is well known to have increased given the parallel between industrialization with associated increased chemical production and utilization and the prevalence of cancer.

Increasing use of chemicals, particularly in the industrializing developing countries (Pakin et al., 1993; Pearce et al., 1994) places new demands on these countries, as they have limited resources to adequately regulate exposure to chemicals. Majority of the chemicals cause mutation in DNA among others. The consequences of increased exposure to chemicals, risk of cancer, early detection of chemical-induced neoplastic changes and the prominent role of biomedical engineering is poorly recognized generally and particularly in the developing countries where chemical carcinogenesis is believed to be currently more prevalent (Huff and Rall,1992). Cancer is classically viewed as the result of series of mutations, including dominantly acting oncogenes and recessively acting tumor- suppressor genes. Each mutation leads to the selective overgrowth of a monoclonal population of tumor cells, and each significant tumor property (invasiveness, metastasis and drug resistance) is accounted for by such mutation (figure 1). The seminal observation that carcinogenesis is a multistage process helps to explain why some chemical carcinogens lack apparently important properties exhibited by others. Such agents may act as promoters on tissues that have been previously initiated or have the ability to produce naturally occurring tumors without treatment.

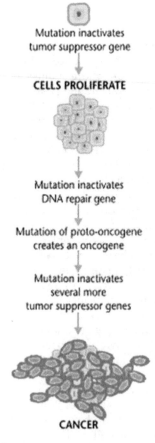

Fig. 1. Stages of the carcinogenic process

DNA-reactive	Activation-independent	Nitrogen mustards, chlorambucil Alkylating agents Epoxides: ethylene oxide
	Activation-dependent	Aliphatic halides: vinyl chloride Aromatic amines: monocyclic-o-toludine; polycyclic-4-aminobiphenyl, benzidine Nitroaromatic compounds: 1-nitropyrene, 3-nitrofluoranthene Heterocyclic amines: 2-amino-3-methylimidazo [4,5-b]pyridine(Ph1P) Aminoazo dyes: dimethylaminoazobenzene Polycyclic aromatic hydrocarbons: benzo(a)pyrene Substituted polycyclic aromatic hydrocarbons: 3-methlycholanthrene N-nitroso compounds: dialkyl-dimethylnitrosamine, diethylnitrosamine; cyclic- N-nitrosonornicotine (NNK), nitrosomorphaline triazines, hydrazines, azoxymethane, methylazoxymethanol, benzene Mycotoxins: aflatoxin B1, aflatoxin G1 Plant products: pyrrolizidine alkaloids, aristolochic acid, cycasin Pharmaceuticals: cyclophosphamide, phenacetin, tamoxifen
Epigenetic	Promoter	Liver enzyme-inducer type hepatocarcinogens: chlordane, DDT, pentachlorophenol, phenobarbital, polybrominated biphenyls, polychlorinated biphenyls Kidney: nitrilotriacetic acid Bladder: sodium saccharin Forestomach-butylated hydroxyanisole
	Endocrine-modifier	Hormones: estrogens-17-estradiol; catechol estrogens-4-hydroxy-estradiol, 2-hydroxyestradiol Estrogen agonists: 17-ethinyl estradiol, diethylstillbestrol (DES) Prolactin inducers: chloro-s-triazines-atrazine Antiandrogens: finasteride, vinclozolin Antithyroid thyroid tumor enhancers Thyroperoxidase inhibitors: amitrole, sulfamethazine Thyroid hormone conjugation enhancers: phenobarbital, Gastrin-elevating inducers of gastric neuroendocrine and glandular tumors: lansoprazole,omeprazole, pantoprazole
	Immunosuppressor	Cyclosporin Purine analogs
	Cytotoxin	Mouse forestomach toxicants: propionic acid, diallyl phthalate, ethyl acrylate Rat nasal toxicants: chloracetanilide herbicides Rat renal toxicants: potassium bromate, nitrilotriacetic acid Male rat kidney _2u-globulin nephropathy inducers: D-limonene, p-dichlorobenzene
	Peroxisome proliferator	Hypolipidemic fibrates: ciprofibrate, clofibrate, gemfibrozil Phthalates: di(2-ethylhexyl)phthalate(DEHP), di(isononyl)phthalate Lactofen
Minerals and Metals		Minerals: asbestos Metals: arsenic, beryllium, cadmium, chromium (IV), nickel, silica
Unclassified		Acrylamide, acrylonitrile, dioxane

Table 1. Classification of chemicals with carcinogenic activity

Foulds (1969) suggested that cancer development consisted of three, rather than two processes: (1) initiation, or the conversion of normal cells to a potentially precancerous form (2) promotion, or the expansion of the clones of initiated cells to form tumors; and (3) progression, or the development of tumors to increasing levels of malignancy. The original view was based on Foulds' wealth of experience with both clinical and experimental cancer. It has however, been expanded greatly since it was first propounded by Foulds (1982). Other investigators have also made significant contributions to the understanding of the process of carcinogenesis by suggesting that there are two major cell-based processes essential to the formation of tumors (Ames and Gold, 1981).The first, or initiating stage, is due to mutation; alteration of the DNA of the affected cell through permanent modification of the DNA. These mutations take place at specific locations on the DNA, referred to as oncogenes and tumor suppressor genes, if these individual cells are to serve as precursors of cancer (Willis, 1960; Klein and Klein, 1984). This area remains intensely investigated in the last couple of decades. What is perhaps worthy of note is that while the activation of an oncogene requires mutation at a specific single base (arrangement of the amines making up the DNA) pair on the DNA template, inhibition of a tumor suppressor gene may be achieved by a much wider range of damaging interactions.

In current research, emphasis is laid on the identification of the genes that are involved in the mutation and subsequent molecular events. The failure in the control mechanisms regulating the expression of and response to tissue growth factors is of considerable interest in chemical carcinogenesis. This contributes to the risk of chemical carcinogenesis and is in turn attributable to a number of factors that will be discussed subsequently. A critical process in carcinogenesis is promotion. This involves cellular proliferation, which involves the division of cells to form two unusually identical cells. This may increase the number of both "normal" and neoplastic mutated or preneoplastic cells, enhancing the chance of a tumor being expressed in a clinically observable form. Surprisingly, such increased levels of cellular proliferation may not be apparent in normal cells of a particular tissue but may occur only in pretumor cells thus making early detection difficult. Tumors are well known to increase in their degree of malignancy with time, a process named "progression" by Foulds. Cohen and Elwein (1991) have suggested that progression is the result of a cascade of further critical mutations in the neoplastic cell population followed by further cell proliferation to increase the number of genetically altered cells and the chance of their forming an increasingly malignant, clinically apparent cancer.

2. Summary of stages of chemical carcinogenesis

Studies indicate that three stages of chemical carcinogenesis can be defined; initiation, promotion and progression as briefly alluded previously.

Initiation: This is concerned with the induction of genetic changes in cells, leading to genome instability. This can be accentuated by micronutrient deficiency disorders. The nature of the initial changes is still incompletely elucidated (Satoh, 1988)

Promotion: This largely involves the induction or commencement of cell proliferation. In this phase of carcinogenesis a promoting agent or enabling microenvironment, brings about increased cell proliferation. This stage is very important as it is reversible if the promoting agents or risk factor(s) are withdrawn. Probably also, if some genome stabilizing micronutrients are abundant. This stage has been exploited considerably for both therapeutic and chemopreventive measures that are in part dependent on micronutrients.

Progression: If cell proliferation is sustained then initiated cells acquire secondary genetic abnormalities in oncogenes which first lead to dysregulation and finally to autonomous growth characteristic of cancer. The ultimate end-point of progression is development of invasive neoplasm.

While many environmental agents can be considered to be chemical carcinogens; some act as both initiators and promoter (complete carcinogens). The understanding of molecular aspects of chemical carcinogenesis has lead to development of the concept of chemoprevention (anticarcinogenesis). This process proposes strategies for intervention at the phase of malignancy using drugs or natural or synthetic agents to reverse or halt the evolution of carcinogenesis which is dependent on recognition of risk factors and early detection.

2.1 Brief history of chemical carcinogenesis

The history of chemical carcinogenesis is punctuated by key epidemiologic observations and animal experiments that identified cancer-causing chemicals and that led to increasingly insightful experiments to establish molecular mechanisms and to reduction of human exposure to chemicals. In 1914, Boveri (1914) made key observations of chromosomal changes, including aneuploidy. His analysis of mitosis in frog cells and his extrapolation to human cancer is an early example of a basic research finding generating an important hypothesis (the somatic mutation hypothesis). The first experimental induction of cancer in rabbits exposed to coal tar was performed in Japan by Yamagiwa and Ichikawa (1918) and was a confirmation of Pott's epidemiologic observation of scrotal cancer in chimney sweeps in the previous century (Potts, 1775). Owing to the fact that coal tar is a complex mixture of chemicals, a search for specific chemical carcinogens was undertaken. British chemists, including Kennaway (1930), took on this challenge and identified polycyclic aromatic hydrocarbons (PAHs), such as, benzopyrene, which was shown to be carcinogenic in mouse skin by Cook and his colleagues in 1933. The fact that benzopyrene and many other carcinogens were polyaromatic hydrocarbons lead the Millers (1947) to postulate and verify that many chemical carcinogens required activation to electrophiles (electron seeking moieties) to form covalent adducts with cellular macromolecules. This in turn prompted Conney and the Millers (1956) to identify microsomal enzymes (P450s) that activated many drugs and chemical carcinogens.

3. Cell regulatory mechanisms: The cell cycle

Carcinogenesis, or the sequence of events leading to cancer, is a multistep process involving both intrinsic and extrinsic factors. In the normal tissue, there are numerous regulatory signals that instruct cells when to replicate and when to die. In a cancer cell these regulatory mechanisms become disabled and the cell is allowed to grow and replicate unchecked. Thus at the most fundamental level cancer is caused by abnormal gene expression. This abnormal gene expression occurs through a number of mechanisms including direct damage to the DNA, and inappropriate transcription and translation of cellular genes. Carcinogenesis has been demonstrated abundantly to be induced or at least caused by exposure to certain types of chemicals (carcinogens). The mechanisms are elaborated on subsequently. The cell cycle plays an important role in this regard. It is concerned with the processes that govern the life and death of cells and through transient delay in G_0 phase or outright apoptosis (programmed cell death) might be able to prevent damage in the DNA of a cell that may proceed to carcinogenesis. In the normal cell, replication of the DNA and cell division is stimulated by the presence of growth factors that bind receptors at the cytoplasmic membrane and initiate a cascade of intracellular signals. Once these signals reach the nucleus they induce transcription of a complex array of genes producing proteins that mediate progression of the cell cycle culminating in mitosis or cell division. One remarkable contribution of biomedical engineering is the introduction of flow cytometer equipment which enable stages of the cell cycle to be followed and disorders or disruptions there of detected.

The cell cycle is conventionally divided into four (4) phases, although there is the subsidiary G_0 phase. The duration of each of these phases varies depending on factors such as cell type and localized conditions within a given tissue (microenvironment). At the end of mitosis (M) daughter cells enter gap 1 (G_1) phase.

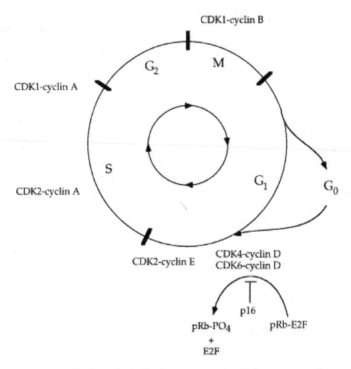

A schematic representation of the mammalian cell cycle. In each cell division cycle, chromosomes are replicated once (DNA synthesis or S-phase) and segregated to create two genetically identical daughter cells (mitosis or M-phase). These events are spaced by intervals of growth and reorganization (gap phases G_1 and G_2). Cells can stop cycling after division, entering a state of quiescence (G_0). Commitment to traverse an entire cycle is made in late G_1. Progress through the cycle is accomplished in part by the regulated activity of numerous CDK–cyclin complexes, indicated here and described in the text.

Fig. 2. Schematic representation of Cell Cycle.

If conditions are favourable cells enter the synthetic (S) phase of the cycle where the entire genome is replicated during DNA synthesis. Following 'S' phase, cells enter the gap 2 (G_2) phase before proceeding through mitosis again. A critical phase boundary exists early in the G_1 phase called the restriction point. This is the point at which the cell must decide to either enter the cell cycle once more or to secondly move into a state of quiescence; G_0 phase. Once committed to this pathway, the cell can either remain in this state of replicative quiescence until it receives a signal to divide again. Alternatively the cell can proceed down a path that leads either to terminal differentiation or to apoptosis. Movement of a cell through the cell cycle is regulated by an enormously complex array of proteins.

The proteins include:

- Cyclins
- Cyclin dependent kinases (CDKs)

- Cyclin activating kinases (CAKs)

- CDK inhibitory proteins.

Binding of an appropriate growth factor at the cell surface starts a signaling cascade that ultimately leads to the expression of the G1 phase cyclins. It is important to remark that in normal cells, external stimuli (factors) such as growth factors are absolutely needed for the cell to proceed beyond the restriction point. Beyond this point, the cell is committed to DNA replication and cell division. Interference with the normal signal transduction pathway by chemical carcinogens, the mechanism notwithstanding can transform a cell into a state of proliferation that is not regulated by normal physiological controls (carcinogenesis). This basically is broadly the molecular basis of carcinogenesis.

Table 2. Proteins Controlling the Cell Cycle.

It is note worthy that all phases of the cell cycle are regulated by the micronutrient zinc. Thus zinc deficiency common in many developing countries (WHO, 2002; Ames, 2010) can be risk factors in chemical carcinogenesis. Ho et al (2003) have elegantly demonstrated this in their studies. This is an area where biomedical engineering has contributed significant in the last five or more decades by the production of flame absorption spectrophotometers (FAAS) and later the graphite furnace (carbon rod) (GFAAS). This was followed by inductively coupled plasma mass spectrometer which allows for simultaneous multi element analysis. These equipment have exquisite sensitivities which enable the status of zinc and many other micronutrients to be detected and indirectly play a preventive role; reducing risk of cancer.

4. Molecular biology and chemical carcinogenesis

The discovery of DNA as the genetic material by Avery, MacLeod, and McCarthy (1944) and the description of the structure of DNA by Watson and Crick (1953) indicated that DNA was the cellular target for activated chemical carcinogens and that mutations (alterations in the sequence of bases making up amino acids) were key to understanding mechanisms of cancer. This led to defining the structure of the principal adducts in DNA (complexes of

DNA and a carcinogen or its metabolite) by benzo (*a*) pyrene (Carrel et al., 1997) and aflatoxin B_1 (Croy et al., 1978). The concepts developed in investigating mechanisms of chemical carcinogenesis also led to discoveries that are relevant to other human conditions in addition to cancer, including atherosclerosis, cirrhosis, and aging. The fact that genetic changes in individual cancer cells are essentially irreversible and that malignant changes are transmitted from one generation of cells to another strongly points to DNA as the critical cellular target modified by environmental chemicals. DNA damage by chemicals occurs randomly; the phenotypes of associated carcinogenic changes are determined by selection.

Epidemiologic studies from all over the world have identified environmental and occupational chemicals as potential carcinogens. The most definitive epidemiologic studies have been those in which a small group is exposed to a tremendously large amount of a specific chemical, such as aniline dyes. The table below (table 3) lists some of the fairly well characterized chemicals sites where they have induced cancer.

Carcinogens		Site of cancer
Chemical mixtures	Soots, tars, oils	Skin, lungs
	Cigarette smoke	Lungs
Industrial chemicals	Benzidine	Urinary bladder
	Nickel compound	Lungs, nasal sinuses
	Arsenic	Skin, Lungs
	Vinyl choride	Liver
Drugs	Mustard gas	Lungs
	Phenacetin	Renal pelvis
Naturally occuring compounds	Cyclomates	Bladder
	Nitroso compounds	Oesophagus, Liver, Kidney, stomach

Table 3. List of Chemical Carcinogens.

5. Mechanisms of chemical carcinogenesis

As part of daily existence, DNA frequently sustains damage. If unrepaired, this can lead to mutations that replicate resulting in abnormal and cancerous development. Some biological mechanisms usually inhibit this process. An enzyme 8-oxoguanine DNA glycosylase (OGG1) among others repairs DNA by excising damaged nitrogen bases constituting the DNA. DNA damage may occur through exposure to chemicals present in cigarette smoke, ionizing radiation and oxidative stress, which can be induced by a number of chemicals such as cadmium and the polycyclic aromatic hydrocarbons. The levels of OGG1 can thus be used to predict an individual's risk of developing cancer.

At least four fundamentally different mechanisms of cancer induction by chemicals have been identified. These may lead to cancer as individual processes or on occasion, the same agent may exert its effects through two or more processes to lead to tumor formation. The importance of the mutation/proliferation approach to the development of cancer lies in its ability to encompass each of these mechanisms within a single frame work. This as has been demonstrated (Clayson, 2001), means that if we can measure changes in mutation and proliferation frequencies, due to a specific carcinogen, there may be no need to elucidate the detailed mechanism of carcinogenesis for every chemical carcinogen before attempting to calculate accurately the risk it may carry for exposed human subjects. This requires exquisite and very sensitive instruments, and appears to be one challenge to specialists in biomedical engineering. Fortunately by the wide range of flow cytometry and mass spectrometry based techniques, the field of biomedical engineering appears to be rising to the challenge. It needs not be emphasized that a great deal of thought and effort will be required if mutation rates and proliferation in specific cell types are to be measured in humans by non-invasiveness.

A number of chemical carcinogens now appear to exert their primary effect on the mutational part of the carcinogenic process, while some others seem to be relatively devoid of the ability to interact with DNA and appear to work mainly through a mechanism of induction of cellular proliferation. Mutation on the other hand appears to be induced by chemical carcinogens by at least two major modes. The modes involve direct interaction with the DNA through the formation of highly reactive, positively charged entities known as "electrophiles" This entity is capable of reacting chemically with many different cellular constituents, including the genetic material, DNA (Miller et al, 1961). The adducts formed with DNA the interaction products of such carcinogen-derived electrophiles with DNA, are not regarded as genetic lesions in their own right. They only represent the first stage in the formation of a mutational event. The adducts may be effectively repaired by the DNA repair enzyme system found in the cell nucleus as earlier indicated. In the alternative, if they are not repaired they may affect important sites on the DNA and consequently die, or if the DNA replicates while they are still present, may lead to mutations through base-mispairing or other errors, that is culminating in true genetic lesions. This has been broadly illustrated by the figure below (figure 3).

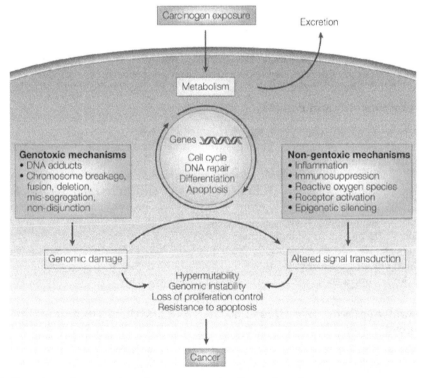

Fig. 3. Overview of genotoxic and non-genotoxic effects of carcinogens.

If the induced mutations occur at one or a relatively few critical sites on the DNA, then the cells may be converted from a "normal" to a preneoplastic state. Chemically-induced mutations are not limited to critical genes, different pretumor cells may demonstrate a variety of different growth potentials due to the range of altered " non-critical" genes, thus enabling those cells with most favourable properties to transform most rapidly to clinically apparent tumours. Alternatively, the carcinogens may act indirectly through the formation of reactive oxygen species or nitrogen radicals, some types of which are also highly reactive with macromolecules such as DNA. The process of raised cellular proliferation is also multi-factorial in its genesis. It may arise from, for instance, direct hormone-like stimulation of specific cell types, from perturbation of tissue processes that lead to a balance between cell proliferation and cell death (apoptosis), it may alternatively arise from massive cell-killing or cytotoxicity followed by proliferative regeneration to maintain the physiological functioning of the affected tissues. A yet further way by which excess, tumor-inducing cellular division may be induced is that exhibited by the urinary bladder. Oyasu and his colleagues (1981) and in his subsequent studies (1995) showed by using heterotopical transplant rat bladder technique that urine by itself, but not water can induce proliferation. The mechanism by which this happens is not quite clear.

It was however conceived that urine contained epithelial growth factors that stimulate cell division and that such factors would penetrate the epithelium should it be injured by the presence of a foreign agent in the bladder. A fourth type of mechanism of carcinogenesis may arise from the ability of the agent to form a complex with a specific protein. This complex (ligand-protein) may have the property of altering the expression of specific and important region in the DNA. There is the emerging but incompletely understood involvement of epigenetics; alteration in the genetic processes not involving DNA base sequence. Many epimutagens have already been identified and a number of existing chemicals such as cadmium are also known to act through this pathway.

6. The environment and cancer

Cancers caused by environmental agents frequently occur in tissues with the greatest surface of exposure to the agents: lung, gastrointestinal tract, and skin. Recently, the study of chemical carcinogenesis has merged with studies on the molecular changes in cancer cells, thus generating biological markers to assess altered metabolic pathways and providing new targets for therapy. Although these are exciting areas, they may be peripheral to attacking the primary causes of the most common human cancers. As more and more mutations are catalogued in cancer cells and more and more changes in transcription regulation, it becomes increasingly apparent that we need to understand what generates these changes. The fact that chemicals cause random changes in our genome immediately implies that our efforts need to be directed to quantifying these changes, reducing exposure, and developing approaches to chemoprevention (Extensively reviewed by Pereira, 1997).

7. Mechanism of oxidative stress and DNA damage due to micronutrient deficiency

Micronutrients are referred to as nutrients required in very small amounts that do not have calorific values but are extremely important for the maintenance of health. They comprise the vitamins and trace elements that regulate vital metabolic and molecular pathways and processes. Some of them play very vital roles in DNA and RNA metabolism either as coenzymes or cofactors involved in their metabolism or as components of systems intimately interacting with these molecular regulators. The micronutrients are basically supplied in the diet or may be taken as supplements. Very few of them are synthesized endogenously enough to meet physiological requirements hence they are mostly essential. Based on public data emanating from the Healthy People 2010 Project, it has been estimated that about 80% of colon and prostate cancers, may be influenced by diet, nutrition, and life style. It has been proposed that DNA damage induced by dietary micronutrient deficiency accounts for about 33% of preventable cancers (Ames, 2001; Ames and Wakimoto, 2002). Owing to the fact that micronutrient deficiencies can induce DNA damage in a manner similar to those induced by ionizing radiation and reactive oxygen species (ROS), it has been suggested that oxidative stress and the associated DNA breaks are critical targets for nutritional control of carcinogenesis (Cheng, 2009) and perhaps a marker for early detection. When DNA lesions are left unrepaired they can promote accumulation of mutations that facilitate the process of carcinogenesis. Micronutrients may act directly on the genome to prevent mutations, or indirectly as enzyme cofactors in cellular processes that modulate transformation (Hanahan and Weinberg, 2000; Sjoblom et al., 2006). By yet incompletely defined mechanisms, micronutrients at levels higher than nutritional requirements may also activate DNA damage response or senescence, which are processes that are recognised to eliminate cancer cells or limit the progression of precancerous cells (Gorgoulis et al., 2005; Bartkova et al., 2005, 2006). The essential microminerals copper, iron, selenium and zinc play important roles in genome stability. In particular, these microminerals have significant impact on oxidative DNA damage and the corresponding repair pathways.

Micronutrient intakes below recommended levels are known to be unusually widespread in poor countries, though also some segments of the population in economically advanced nations such as the United States, especially among the poor, children, adolescents (Anetor, 2009; Ames, 2010). It has been hypothesised that two of the many insidious but measurable consequences of moderate micronutrient inadequacy are increased DNA damage (precursor of cancer) and mitochondria decay which can cause mutagenic oxidant release also involved in future carcinogenesis. Studies indicate that sensitive assays targeted at these end points have a high probability of detecting changes in individuals with micronutrient deficiencies(Ames, 2010) This again appears instructive to scientists and technologists in the biomedical engineering field.

8. Technology and chemical carcinogenesis

Many technological advances have allowed conceptual ideas to be experimentally tested, including the sensitive detection of chemical carcinogens by high-pressure liquid chromatography (Esaka, et al., 2003) and mass spectrometry (Sigh and Farmer,2006), detection of DNA adducts by postlabeling (Randerath et al, 1981) and by specific antibodies (Poirier et al., 1977), transcriptional profiling by arrays (Kallioniemi et al., 1992, Schena et al., 1995), and quantitation of mutagenicity of carcinogens using bacterial genetics (Ames et al., 1973). The testing of certain concepts in chemical carcinogenesis awaited the development of new technologies. For example, the concept of somatic mutations in cancer preceded by 40 years the establishment of DNA as the genetic material and by 67 years the development of DNA sequencing methods that directly showed clonal mutations in human cancer cells. Also, the mutator phenotype hypothesis formulated in 1974 has been only recently experimentally verified. The list below shows some commonly employed analytical techniques requiring the attention of biomedical engineering and technology.

Some analytical areas that may be particularly relevant in chemical carcinogenesis will include:

- 32P-post labelling
- Fluorescent-based techniques
- Immunoassay-based techniques
- Electrochemical detectors
- Electron microscopy
- High performance liquid chromatography/ Ultra high performance liquid chromatography
- Comet (Single cell gel electrophoresis), imaging system
- Mass spectrometry
- Atomic absorption spectrophotometry (AAS)
- Inductively coupled plasma- mass spectrometry (ICP-MS)
- Advanced spectrophotometry
- DNA microarray systems

The latter is particularly promising as it enables the simultaneous measurement of transcription of thousands of genes using microchips containing thousands of probes of complementary DNA (cDNA) immobilized in predetermined array. But suffers the caveat of being very expensive especially for the developing countries that appear to be mostly in need of it currently.

9. Early studies in chemical carcinogenesis and risk factors

Early in the field of chemical carcinogenesis, investigators recognized that perturbation of the normal microenvironment by physical means, such as wounding of mouse skin or partial hepatectomy in rodents (Hennings and Boutwell, 1970; Fausto etal, 2006) or chemical agents, such as exposure of the mouse skin to certain phorbol esters (Berenblum, 1941), can drive clonal expansion of the initiated cells toward cancer. In the second stage, tumor promotion results in proliferation of the initiated cells to a greater extent than normal cells and enhances the probability of additional genetic damage, including endogenous mutations that accumulate in the expanding population. This classic view of two-stage carcinogenesis (Berenblum, 1941) has been conceptually important but also an oversimplification of the increasing understanding of the multiplicity of biological processes that are deregulated in cancer. In addition, an active debate continues on the relative contribution of procarcinogenic endogenous mechanisms—for example, free-radical-induced DNA damage (Halliwell and Aruoma, 1991), DNA depurination (Lindahl and Nyberg, 1972), DNA polymerase infidelity (Loeb et al, 1974), and deamination of 5-methylycytosine (Lindahl and Nyberg, 1974)—compared with exposure to exogenous environmental carcinogens (Ames et al, 1973).

The enhancement of carcinogens by epigenetic mechanisms such as halogenated organic chemicals and phytoestrogens (Martin et al, 2007), as well as the extrapolation of results from animal bioassays for identifying carcinogens to human cancer risk assessment, are also difficult to quantify (Swenberg et al., 1998). As discussed below, this debate is not merely an academic event, in that societal and regulatory decisions critical to public health are at issue. The identification of chemical carcinogens in the environment and occupational settings [benzo (a) pyrene and tobacco-specific nitrosamines in cigarette smoke, aflatoxin B_1 (AFB_1), residues from fossil fuel, vinyl chloride, pesticides and benzene] has led to regulations that have reduced the incidence of cancer. Further reduction or near total elimination may be achieved by sensitive instruments that enable early detection of up- stream changes that may culminate in cancer. This needless to say has heavy reliance on biomedical engineering.

9.1 Risk factors and early detection: role for micronutrient deficiency and oxidative stress

The risk of contracting cancer generally increases as the population grows older; this has been reported to be directly proportional to the number of years raised to the fourth power (Tolonen, 1990). This may be modified in the case of chemical carcinogenesis to include dose of chemicals or environmental agent to which the population is exposed and the stoichiometric bioavailability of protective factors such as the micronutrients. In animal models, such as the mice, the life-span of the models has been increased and cancer prevented by calorie restriction (reducing oxygen intermediates) and feeding them antioxidants. This suggests that excess calorie and antioxidant deficit (oxidative stress) are risk factors that may enhance the carcinogenic process. The natural dietary antioxidants, selenium, zinc, vitamins A, C, and E plus β-carotene protect against free radicals, lipid peroxidation (Tolonen, 1990) and thus the risk of chemical carcinogenesis. Ames (1983) has greatly emphasised the role of antioxidants largely derived from micronutrients as anticarcinogenesis. Vitamin C for instance is well known to counteract carcinogenic nitrous amine in the stomach. Urban population are more exposed to the risk of cancer than rural dwellers. This perhaps can be explained by the probability that the rural population is exposed to fewer carcinogens present in the environment.

Additionally, there is the often neglected element of greater host resistance in that rural populations are more likely to consume diet replete with antioxidant micronutrients some of which will also enhance the immune system. This is important in that its contact with carcinogens either in the diet or environment is inevitable. Thus it may be possible for us to avoid the most prominent risk factors the most pragmatic option appears to be reinforcing host resistance .This may be enhanced by the use of biomarkers. Biomarkers are playing an increasing role in the assessment of human exposure to hazardous environmental pollutants or chemicals and in risk assessment to these compounds. Biomarkers may be applied at any stage in the toxicological process, ranging from measurement of the external dose as an indicator of exposure to determine altered structure and function of cells as a marker of effect- carcinogenesis. Genetic carcinogens interact with nucleic acids to produce adducts, measurement of which is an indicator of the dose of active material which has reached the cells in question, termed biologically active dose (BAD), in the individual. This consequently incorporates the inter-individual variation in absorption, metabolism, and excretion of the compound which may affect risk assessment.

10. DNA repair: Introduction

To maintain the genomes of organisms, they have evolved a network of DNA repair pathways to excise altered residues from DNA. A major consideration is the relative contribution of environmental and endogenous DNA damage to carcinogenesis. DNA damage by environmental agents would have to be extensive and exceed that produced by normal endogenous reactive chemicals to be a major contributor to mutations and cancer.

This consideration underlines the difficulty in extrapolating risk of exposure to that which would occur at very low doses of carcinogens

10.1 DNA repair and role of micronutrients as biomarkers of susceptibility to DNA damage

Very many factors including nutritional factors have been shown to delay the carcinogenic process. Thus this can be exploited to reverse, delay or prevent the carcinogenic process. Maintenance of genome stability is of fundamental importance for counteracting carcinogenesis. Many human genome instability syndromes exhibit a predisposition to cancer. An increasing body of epidemiological evidence has suggested a link between nutrient status and risk of cancer. Populations in developing countries that are deficient in these protective micronutrients (WHO, 2002; Ames, 2010) and are increasingly exposed to chemicals owing to progressive or rapid industrialization are thus at increased risk (Anetor et al., 2008). Based on public data from the healthy people 2010 project, it is estimated that up to 80% of colon and prostate cancers may be influenced by diet, nutrition and life styles. As earlier indicated, it has been proposed that DNA damage induced by dietary micronutrient deficiency accounts for one-third of preventable cancers. Because micronutrient deficiencies can induce DNA damage in forms similar to those induced by ionizing radiation and reactive oxygen species (ROS), it has been suggested that oxidative stress and associated DNA breaks are critical targets for nutritional control of carcinogenesis. If left unrepaired, DNA lesions can promote accumulation of mutations that facilitate the process of carcinogenesis.

Micronutrients may act directly on the genome to prevent mutations, or indirectly as enzyme cofactors in cellular processes that modulate transformation. Thus micronutrient status may serve as biomarkers of risk of carcinogenesis. For instance low selenium status is a biomarker of risk of many cancers including cancer of the prostate. This should be particularly appealing to industrializing developing countries. Human cells possess an armamentarium of mechanisms for DNA repair that counter the extensiveness of DNA damage caused both by endogenous and environmental chemicals. These mechanisms include base excision repair (BER) that removes products of alkylation and oxidation (Duncan et al, 1976; Roth and Samson, 2002; Gersson, 2002); nucleotide excision repair (NER) that excises oligonucleotide segments containing larger adducts (Setlow and Carrier, 1963); mismatch repair that scans DNA immediately after polymerization for misincorporation by DNA polymerases (Modrich, 1991); and oxidative demethylation (Sedgwick, 2004), transcription-coupled repair (TCR) that preferentially repairs lesions that block transcription (Hanawalt, 1994); double-strand break repair and recombination that avoids errors by copying the opposite DNA strand (Friedberg et al, 2005); as well as mechanisms for the repair of cross-links between strands (Kuraoka et al, 2000; Zheng et al, 2005) that yet need to be established. Micronutrients deficiency disorder may inhibit DNA repair, thus acting as risk factors. Determinations of the levels of micronutrients may therefore serve as biomarker of susceptibility to DNA damage using the various instruments provided by biomedical engineering. Micronutrient deficiency for instance is inversely correlated with the level of 8-hydrodeoxyguanosine (8-OHdG), a marker of oxidative DNA damage, which is mutagenic and has to be removed by protective enzymes such as the human oxo-guanine DNA glycosylase (hOGG1).

Most DNA lesions are subject to repair by more than one pathway. As a result, only a minute fraction of DNA lesions which escapes correction are present at the time of DNA replication and can direct the incorporation of noncomplementary nucleotides resulting in mutation. Unrepaired DNA lesions initiate mutagenesis by stalling DNA replication forks or are copied over by error-prone *trans*-lesion DNA polymerases (McCulloch and Kunkel, 2008). Alternatively, incomplete DNA repair can result in the accumulation of mutations and mutagenic lesions, such as abasic sites (Loeb, 1985). Maintenance of genome stability is crucial for avoiding carcinogenesis. A number of human cancers display a range of chromosomal abnormalities; a characteristic now termed genome instability. The relationship between cancer and genome instability is well recognized, but the causes of genome instability in the evolution of human cancers is incompletely elucidated. The DNA damage response safeguards the integrity of the genome by detecting alterations, halting cell cycle progression and repairing damaged DNA. Zinc which plays a role in all the phases of cell cycle when deficient can be critical. (Anetor et al, 2008)   Cells with defective DNA damage responses are characterized by high level of genome instability (Cheng, 2009).

It is known that in particular, cells in s-phase are vulnerable to agents, such as chemicals in the environment that cause DNA damage and induce DNA replication fork arrest. Since such events can adversely affect genomic stability, cells have evolved S-phase DNA response cascades, including checkpoint responses and DNA repair mechanisms, to fix DNA damage (Bartek et al., 2004). In response to DNA damage, check points are activated that coordinate DNA damage signaling, cell cycle arrest and DNA repair. Cells have developed elaborate systems to repair varieties of DNA damage. Abundant evidence has linked defects in DNA repair to carcinogenesis. As a way of avoiding mutagenic events, the DNA base excision repair (BER) pathway copes with oxidatively modified DNA (Xu et al., 1997; Kungland et al., 1999), and nucleotide excision repair can deal with bulky DNA adducts including DNA cross-links (Cleaver, 2005). It is noteworthy that micronutrient deficiency significantly affects DNA damage repair. Zinc (Zn) deficiency is common in children and adults (WHO, 2002; Moshfegh et al., 2005). Human cell culture studies demonstrating severe Zn deficiency causes complex IV deficiency and the release of oxidants, resulting in significant oxidative DNA damage (Ho and Ames, 2002). Zinc deficiency has also been reported to cause chromosome breaks in rats (Bell et al, 1975) which has been associated with cancer in both animal models and humans (Fong et al., 2005).

These reports strengthen the significant effect of micronutrient deficiency on DNA damage repair and by extrapolation on risk of carcinogenesis particularly chemical carcinogenesis in populations exposed to chemicals. Zinc deficiency in human cells has also been shown to inactivate Zn-containing proteins such as the tumor suppressor protein, p53 which plays a significant role in genome protection (Lane, 1992) and the DNA base excision repair enzyme,apyrimidinic/apurinic endonuclease, with a resulting synergistic effect on genetic damage (Ho and Ames, 2002;Ho Courtemanche and Ames, 2003).

11. Integrative cell biology and chemical carcinogenesis

Damage to DNA by chemical carcinogens activates checkpoint signaling pathways leading to cell cycle arrest and allows time for DNA repair processes (Sweasy et al, 2006). In the absence of repair, cells can use special DNA polymerases that copy past DNA adducts (Masutani et al., 2000) or undergo apoptosis by signaling the recruitment of immunologic and inflammatory host defense mechanisms. The demonstration that each methylcholanthrene-induced tumor has a unique antigenic signature provided one of the earliest glimpses into the stochastic nature of cellular responses to carcinogens. The immunologic and inflammatory responses facilitate not only engulfment and clearance of damaged cells but also the resulting generation of reactive oxygen (Klebanoff, 1988) and nitrogen radicals (Ohshima and Batsch, 1994) that further damage cellular DNA.

12. Chemical carcinogens and induced somatic mutations as biomarkers in molecular epidemiology

Extensive experience of laboratory research in chemical carcinogenesis has provided a solid foundation for the analysis of chemical-specific macromolecular adducts and related somatic mutations in humans as biomarkers of carcinogen exposure. A paradigm for validating causal relationships between biomarkers of carcinogens exposure and a cancer risk biomarker is shown in a number of observations (Wogan et al, 2005). The metabolite, AFB_1, a fungal toxin, is a prototypical example of an environmental chemical carcinogen that has been validated using this strategy. Benzo (a) pyrene, a polycyclic aromatic hydrocarbon 4-aminobiphenyl (Vineis and Pirastu, 1997), an aromatic amine dye, and 4-(N-methyl-N-nitrosamino)-1-(3-pyridyl)-1-butanone, a tobacco-specific N-nitrosamine (Hetch, 1999), are other key examples. The prevalence of Aflatoxin B1 in developing countries makes host resistance mechanism based on micronutrients of great import to reduction of risk of chemical carcinogenesis.

13. Impact of new technologies or analytical techniques and biomedical engineering

Recent advances in molecular methodologies are phenomenal, and they increasingly are being applied to understanding the interaction of chemical carcinogens with cellular constituents and metabolism. Cloning of DNA has facilitated the identification of specific genes mutated in human cancers. Chemical methods, including mass spectrometry, allow us to measure carcinogen alteration with unprecedented sensitivity and specificity, particularly the ultra high performance liquid chromatography (UHPLC) coupled to MS. Mass spectrometry is being coupled with many other site-specific techniques to study mutagenesis; to define how specific alterations in DNA produce cognate mutations. Sequencing of the human genome and the identification of DNA restriction enzymes opens up the field of molecular epidemiology, focusing in part on individual susceptibility to carcinogens.

A very useful technique in cancer studies is the single cell protein electrophoresis (COMET ASSAY). This assay is the only direct method for the detection of DNA damage in cells. It is used in cancer research, in genotoxicity studies on environmental mutagens, and for screening compounds for cancer therapeutics.

DNA microarray technology is a powerful tool that allows the activity of several genes to be monitored simultaneously. DNA microarray has been especially useful for detecting genes that respond to a specific chemical or physical signal in the same way. A DNA microarray or DNA chip consists of a solid surface, usually glass, to which DNA fragments are attached. The copies of each kind of fragments are attached to the glass surface at a specific site to regular pattern or array. DNA fragments attached to the chip act as probes that can hybridize with complimentary DNA or RNA molecules (targets) in solution. DNA microarray technology has been used to study a wide variety of gene expression problems such as tissue specific, cell cycle specific, and tumor- specific gene activation and repression.

Once a similar pattern expression profile has been established for a group of genes, it seems reasonable to assume that the profile is at least partly caused by similar transcription regions. Therefore, if information is available about a regulatory region in one gene within this group, it may provide clues to the regulatory regions of other members of the group as well as to the protein factors that activate or inhibit gene expression. The initial stimulus may be a given carcinogen. Array technology facilitates analysis of carcinogen-induced alterations in the expression of both protein coding and noncoding genes. These are all areas where biomedical engineering is expected to play a pivotal role in risk identification and early detection. On the horizon are techniques that can measure single molecules of carcinogens in cells, random mutations in individual cells, analysis of the dynamics of how molecules exist and work, and bioinformatics and genetic maps to delineate complex interacting functional pathways in cells. Underlying this progress in understanding chemical carcinogenesis is a cascade of advances in molecular biology that makes it feasible to quantify DNA damage by chemical agents, mutations, and changes in gene expression.

This calls for very intimate collaboration between biotechnology and biomedical engineering, a partnership that promises to be very rewarding in the fight against chemical carcinogenesis. Determining the structure of DNA, DNA sequencing, and the PCR revolutionized cell biology, including carcinogenesis. Advances in detection of DNA damage, including postlabeling of DNA (Randerath et al, 1981), immunoassays (Poirier et al, 1977) and mass spectrometry as earlier discussed (Singh and Farmer, 2006) have allowed the detection of a single altered base in 10^9 nucleotides using human nuclear DNA. This technology can be extended to analyze DNA or RNA in a single cell (Klein, 2005). Advances in cell biology, including array technology (Schena et al, 1995) and proteomics (Anderson et al, 1984; Aebersold et al, 1987), make it feasible to assess global changes in RNA and protein expression during carcinogenesis. Together, these technologies underlie systems biology, making it increasingly feasible to map biochemical pathways in cancer cells from DNA, to RNA, to proteins, to function. This again calls for greater involvement of the biomedical engineering field.

The ultimate application of techniques of biomedical engineering in the field of chemical carcinogenesis holds great promise, but faces several formidable challenges requiring ingenuity in matching technology with a social obligation to ensure that those most affected can afford its dividend to combat the menace of chemical carcinogenesis. This calls for sensitive methods for the early detection. The complex interplay among these and the great promise they hold for human health, particularly in the rapidly industrializing developing countries has not been adequately addressed. This contribution largely sees this as needing to be addressed- carcinogenesis, integrating chemical exposure, nutritional; mainly micronutrient modulation and the yawning gap biomedical engineering should fill in non-invasive applications. Although the field of biomedical engineering as regards cancer studies is still relatively in its early stages and will be a long-term effort and the magnitude of the task far greater than the physical resources and intellectual capacity currently available, the challenge is to focus on sensitive, specific or selective, cost effective and efficient techniques with dereliction for resource poor rapidly industrializing developing countries that appear to bear a greater brunt of the increasing chemical exposure and attendant carcinogenic risk.

14. References

Aebersold R. H., Leavitt. J., Saavedra, R. A, Hood, L. E. and Kent, S. B. (1987). Internal amino acid sequence analysis of proteins separated by one- or two-dimensional gel electrophoresis after in situ protease digestionon nitrocellulose. *Proc. Natl. Acad. Si. USA*. 84:6970-6974

Ames, B. N. and Wakimoto, P. (2002). Are vitamin and mineral deficiencies a major cancer risk? *Nat. Rev. Cancer*. 2:694-704.

Ames, B. N. (2001). DNA damage from micronutrient deficiencies is likely to be a major cause of cancer, *Mutat. Res*. 475:7-20.

Ames, B.N., Gold, L. S. and Willet, W. C. (1995). The causes and prevention of cancer. *Proc. Natl. Acad. Sci*. USA 92: 5258–5265.

Ames, B. N. and Gould, C. S. (1991). Endogenous mutagenesis and the causes of aging and cancer. *Mutations Research*, 250: 3.

Ames, B. N. (1983). Dietary carcinogens and anticarcinogenesis. *Science*. 221:1256-1264.

Ames, B. N. Durston, W. E., Yamasaki, E. and Lee, F.D. (1973). Carcinogens are mutagens: A simple test system combining liver homogenates for activation and bacteria for detection. *Proc. Natl. Acad. Sci*. USA. 70: 2281-2285.

Anderson, N. L., Hofmann, J. P., Gemmell, A. and Taylor, J. (1984). Global approaches to quantitative analysis of gene-expression patterns observed by use of two-dimensional gel electrophoresis. *Clin. Chem.*, 30: 2031-2036.

Anetor J. I, Anetor G. O, Udah,D.C, Adeniyi F. A. A(2008): Chemical carcinogesis and chemoprevention:Scientific priority area in rapidly industrializing developing countries. African Journal of Environ. Sci. Tech. 2.(7):150-156

Avery, O. T., MacLeod, C. M. and McCarty M.(1944). Studies on the chemical nature of the substance inducing transformation of pneumoccal types. Induction of transformation by a desoxyribonucleic acid fraction isolated from pneumoccus type III. *J Exp Med*.79:137–58.

Bartek, J., Lukas, C and Lukas, J. (2004). Checking on DNA damage in S phase. *Nat. Rev. Mol. Cell Biol*.5:792-804.

Bartkova, J., Rezaei, N., Liontos, M., Karakaidos, P et al., (2006). Oncogene-induced senescence is part of the tumorigenesis barrier imposed by DNA damage checkpoints. *Nature*. 444:633-637.

Bartkova, J., Horejsi, Z., Koed, K., Kramer,A.,et al., (2005). DNA damage response as a candidate anticancer barrier in early human tumorigenesis. *Nature*. 434:864-870.

Bell, L. T., Branstrator, M., Roux, C., and Hurley, L. S. (1975). Chromosomal abnormalities in maternal and fetal tissues of magnesium or zinc deficient rats. *Teratology*. 12(3):221-226.

Berenblum, I. (1941). The mechanism of carcinogenesis. A study of the significance of cocarcinogenic action and related phenomena. *Cancer Research*, 1: 807.

Block, G., B. Patterson, B. and Subar, A. (1992). Fruit, vegetables and cancer prevention: A review of the epidemiologic evidence. *Nutr. Cancer* 18: 1–29.

Burcham, P. C. (1998). Genotoxic lipid peroxidation products: Their DNA damaging properties and role in formation of endogenous DNA adducts. *Mutagenesis* 13: 287–305.

Cairns, J. (1981). The origin of human cancers. *Nature*, 289: 353-357.

Carrell, J. C., Carrell, T. G., Carrell, H. L., Prout, K. and Glusker, P.(1997). Benzo[a]pyrene and its analogues: structural studies of molecular strain. *Carcinogenesis*, 18:415–22.

Cheng, W. H. (2009). Impact of inorganic nutrients on maintenance of genomic stability. *Environ. Mol. Mutagenesis*. 50:349-360.

Clayson, D.B. (2001).*Toxicological carcinogenesis*. Lewis Publishers, Florida.

Cleaver, J. E. (2005). Cancer in xeroderma pigmentosum and related disorders of DNA repair. *Nat. Rev. Cancer.* 5:564-573.

Cohens, S. M. and El Wein, C. B. (1991). Genetic erors, cell proliferation and carcinogenesis. *Cancer Research*, 51: 6493.

Croy, R. G., Essigman, J. M., Reinhold, V. N. and Wogan, G. N.(1978). Identification of the Principal Aflatoxin B1-DNA Adduct Formed *in vivo* in Rat Liver. *PNAS*, 75:1745-9.

Duncan, J., Hamilton, L. and Friedberg, E. C. (1976). Enzymatic degradation of uracil-containing DNA. *J. Virology*, 19: 338-345.

Eden, A., Gaudet, F., Waghmare, A. and Jaenisch, R.(2003). Chromosomal instability and tumors promoted by DNA hypomethylation, *Science* 300:455.

Esaka, Y., Inagaki, S. and Goto, M. (2003). Seperation procedures capable of revealing DNA adducts. *J. Chromatography B.*, 797: 321-329.

Fausto, N., Camobekk, J. S. and Riehle, K. J. (2006). Liver regeneration. *Hepatology*, 43: S45-53.

Fenech, M. (2001). The role of folic acid and Vitamin B12 in genomic stability of human cells, *Mutat. Res.* 475: 57–67.

Fong, L. Y. Y., Zhang, L., Jiang, Y., and Farber, J. L. (2005). Dietary Zinc modulation of COX-2 expression and lingual and esophageal carcinogenesis in rats. *J. Natl. Cancer Instit.* 97(1):40-50.

Foulds, L. (1961). Neoplastic development. Vol. 1. Academic press, London.

Friedberg, E. C., Walker, G. C., Siede, W., Wood, R. D., Schultz, R. A. and Ellenberger, T. (2005). DNA repair and mutagenesis 2. ASM press, Washington D. C.

Gerson, S. L. (2002). Clinical relevance of MGMT in the treatment of cancer. *J. Clinical Oncology*, 20: 2388-2399.

Gerster, H. (1995). Beta-carotene, vitamin E and vitamin C in different stages of experimental carcinogenesis. *Eur. J. Clin. Nutr.* 49: 155–168.

Gorgoulis V. G.,Vassiliou, L. F., Karakaidos, P., Zacharatos, P., et al., (2005). Activation of the DNA damge checkpoint and genomic instability in human precancerous lesions. *Nature.* 434:907-913.

Hagen, T. M., Ingersoll, R. T., Liu. J., Lykkesfeldt. J., Wehr. C. M., Vinarsky, V., Barthelomew, J. C. and Ames, B. N. (1998). (R)-α-Lipoic acid-supplemented old rats have improved mitochondrial function, decreased oxidative damage, and increased metabolic rate. *FASEB J.* 13: 411–418.

Hanahan, D. and Weinberg, R. A. (2000). The hallmarks of cancer. *Cell*.100:57-70.

Halliwell, B. and Aruoma, O.I. (1991). DNA damage by oxygen-derived species. Its mechanism and measurement in mammalian systems. *FEBS Lett.*, 291: 9-19.

Hanawalt, P. C. (1994). Transcription-coupled repair and human disease. *Science*, 266: 1957-1958.

Hecht, S.S. (1999). Tobacco smoke carcinogens and lung cancer. *J. Natl. Cancer Inst.* 91: 1194-1210.

Henning, S. M., Swendseid, M. E. and Coulson, W. F. (1997). Male rats fed methyland folate-deficient diets with or without niacin develop hepatic carcinomas associated with decreased tissue NAD concentrations and altered poly(ADP-ribose) polymerase activity. *J. Nutr.* 127: 30–36.

Hennings, H. and Boutwell, R. K. (1970). Studies on the mechanism of skin tumor promotion. *Cancer Research,* 30:312-20.

Ho, E and Ames, B.N. (2002). Low intracellular zinc induces oxidative DNA damage, disrupts p53, NFkappa B, AP1 DNA binding and affects DNA repair in a rat glioma cell line. *Proc. Natl. Acad. Sci. USA.* 99:16770-16775.

Ho, E., Courtemanche, C., and Ames, B. N. (2003). Zinc deficiency induces oxidative DNA damage and increases p53 expression in human lung fibroblast. *J. Nutr.* 133:2543-2548.

International Agency for Research on Cancer. (1971-2006). IARC monographs on the evaluation of carcinogenic risks to humans. Overall evaluation of carcinogenicity. Monographs Volumes 1 to 99. http://monographs.iarc.fr.

Kallioniemi A., Kallioniemi, O. P., Sudar, D., et al., (1992) Comparative genomic hybridization for molecular cytogenetic analysis of solid tumors. *Science.* 258:818-821.

Klebanoff, S. J. (1988). Phagocytic cells: Products of oxygen metabolism. In: Gallin, J. I., Goldstein, I. M. and Snyderman, R. (editors). Inflammation: Basic principles and clinical correlates. Raven press Ltd, New York. Pp391-444.

Klein, C. A. (2005). Single cell amplification methods for the study cancer and cellular ageing. *Mech. Ageing Dev.,* 126: 147-151.

Klein, G. (1987). The approaching era of the tumor suppressor genes. *Science,* 238: 1539.

Klein, G. and Klein, E. (1984). Commentry: Oncogene activation and tumor progression. *Carcinogenesis,* 5: 429.

Klungland, A., Rosewell, I., Hollenbach, S., Larsen, E., et al., (1999). Accumulation of premutagenic DNA lesions in mice defective in removal of oxidative base damage. Proc. *Natl. Acad. Sci. USA.* 96:13300-13305.

Kuraoka, I., Kobertz, W. R., Ariza, R.R., Biggerstaff, M., Essigmann, J. M. and Wood, R. D. (2000). Repair of an interstrand DNA cross-link initiated by ERCC1-XPF repair/recombination nuclease. *J. Biol. Chem.,* 275: 26632-26636.

Lindahl, T. and Nyberg, B. (1974). Heat-induced deamination of cytosine esidues in deoxyribonucleic acid. Biochem. 13:3405-3410.

Lindahl, T. and Nyberg, B. (1972). Rate of depurination of native deoxyribonucleic acid. *Biochemistry.* 11: 3610-3618.

Loeb, L. A. (1985). Apurinic sites as mutagenic intermediates. *Cell.* 40:483-484

Martin, J. H., Crotty, S. and Nelson, P. N. (2007). Phytoestrogens: Perpetrators or protectors? *Future Oncology,* 4: 3007.

Masutani, C., Kusumoto, R., Iwai, S. and Hanaoka, F. (2000). Mechanisms of accurate translesion synthesis by human DNA polymerase eta. *EMBO J.,* 19: 3100-3109.

McCulloch, S. D. And Kunkel, T. A. (2008). The fidelity of DNA synthesis by eukaryotic replicative and translesion synthesis polymerases. Cell Res. 18:148-161

Miller, E.C., Miller, J. A. and Hartman, J. A. (1961). N-Hydroxy-2-acetyllaminofluorine: A metabolite of 2-actylaminofluorine with increased carcinogenic activity in the rat carcinogenesis. *Cancer Research,* 2:815.

Modrich, P. (1991). Mechanisms and biological effects of mismatch repair. *Annu. Rev. Genet.,* 25: 229-253.

Moshfegh, A., Goldman,J., and Cleveland, L. (2005). "What we eat in America. NHANES 2001-2002: Usual nutrient intakes from food compared to dietary reference intakes," U. S. Department of Agriculture, Agricultural Research Service.

Ohshima, H. and Bartsch, H. (1994). Chronic infections and inflammatory processes as cancer risk factors: possible role of nitric oxide in carcinogenesis. *Mutat. Res.* 305: 253-264.

Oyasu, R. (1995). Epithelial tumors of the lower urinary tract in humans and rodents. Food and *Chemical Toxicology*, 33: 747.

Oyasu, R. Hirao, I. and Izumi, K. (1981). Enhancement cement by urine of urinary bladder carcinogenesis. *Cancer Research*, 41: 478.

Perera, F. P.(1997): Environment and Cancer: Who are susceptible?. Science vol. 278:1068-1073.

Poirier, M. C., Yuspa, S. H., Weinstein, I. B. and Blobstein, S. (1977). Detection of carcinogen-DNA adducts by radioimmunoassay. *Nature*, 270: 186-188.

Randerath, K., Reddy, M. V. and Gupta, R. C. (1981). 32P-labelling test for DNA damage. *Proc. Natl. Acad. Sci. USA.*, 78: 6126-6129.

Roth, R. B. and Samson, L.D. (2002). 3-Methyladenine DNA glycosylase-deficient Aag null mice display unexpected bone marrow akylation resistance. *Cancer Research*, 62: 656-660.

Schena, M., Shalon, D., Davies, R. W. and Brown, P. O. (1995). Quantitative monitoring of gene expression patterns with a complementary DNA microarray. *Science*, 270: 467-470.

Sedgwick, B. (2004). Repairing DNA-methylation damage. *Nat. Rev. Mol. Cell Biol.*, 5: 148-157.

Setlow, R. B. and Carrier, W. L. (1963). The disappearance of thymine dimmers from DNA: An error-correcting mechanism. *Proc. Natl. Acad. Sci.* USA. 51: 226-231.

Singh. R. and Farmer, B.P. (2006). Liquid chromatography-electro-spray-ionization-mass spectrometry: The future of DNA adduct detection. *Carcinogenesis*, 27: 178-196.

Sjoblom T., Jones. S., Wood, L. D. et al. (2006). The concensus coding sequences of humanbreast and colorectal cancers. *Science.* 314: 268-274.

Solt, D. and Farber, E. (1976). New principle for the analysis of chemical carcinogenesis. *Nature*, 263:70b

Sweasy, J. B., Lauper, J. M. and Eckert, K. A. (2006). DNA polymerases and human diseases. *Radiat. Res.* 166: 693-714.

Swenberg, J.A., Richardson, F.C., Boucheron, J. A. et al., (1987). High- to low-dose extrapolation: Critical determinants involved in the dose response of carcinogenic substances. *Environmental Health Perspectives*, 76: 57-63.

Tolonen, M. (1990). *Vitamins and minerals in health and nutrition.* Ellis Horwood Ltd. Chichester , West Sussex, England.

Vineis, P. and Pirastu, R. (1997). Aromatic amines and cancer. *Cancer Causes Control*, 8: 346-355.

Watson, J. D. and Crick, F. H.(1953). Molecular structure of nucleic acids: a structure for deoxyribose nucleic acid.*Nature*, 171:737–8.

Weinberg, E (1981). Review: Iron and neoplasia. *Biol. Trace Element Res.* 3:55 - 80.

Weinberg, E. (1984). Iron withholding: A defense against infection and neoplasia. *Physiol. Rev.* 64(1):65- 102.

WHO (2002). The world health report 2002. Reducing risk, promoting healthy lifestyle. World Health Organisation, Geneva.

Willett, W.C. and Trichopoulos, D.(1996). Nutrition and cancer: A summary of the evidence. *Cancer Causes and Control* 7: 178–180.

Wogan, G. N., Hecht, S. S., Felton, J. S., Conney, A. H. and Loeb, L. A. (2004). Environmental and chemical carcinogenesis. *Semin. Cancer Biol.*, 14: 473-486.

Xu, Y., Moore, D. H., Broshears,J., Liu, L., Wilson, T. M. and Kelley M.R. (1997). The apurinic/apyrimidinic endonuclease (APE/ref-1) DNA repair enzyme is elevated in premalignant and malignant cervical cancer. Anticancr Res. 17:3713-3719.

Yamagiwa, K. and Ichikawa, K. (1918). Experimental study of the pathogenesis of carcinoma. J. Cancer Research, 3:1-21.

Zheng, H., Wang, X., Legerski, R. J., Glazer, P. M. and Li, L. (2006). Repair of DNA interstrand cross-links: Interactions between homology-dependent and homology-independent pathways. *DNA Repair*, 5: 566-574.

3

Cell Signalling and Pathways Explained in Relation to Music and Musicians

John T. Hancock

University of the West of England, Bristol, UK

1. Introduction

Cell signalling is arguably the most important area of modern biology. The subject encompasses the control of cellular events, especially in response to extracellular factors. It has been suggested that the human body is one of the most complex machines ever produced (Dawkins, 1989) and the regulation of the activities within it are also equally complex.

Interest in cell signalling does not simply stem from an academic viewpoint either. Certainly there is a vast resource of research which is focuses on the investigation of signalling pathways and the control which they bestow on a cell. However, there are tangible reasons to take an interest here too. The vast majority of new pharmaceutical compounds under development are aimed at the modulation of proteins involved in cell signalling events (Filmore, 2004). Such proteins may be G protein-coupled receptors (GPCRs) or perhaps kinases which are downstream of such receptors. Many anti-cancer studies are now focused on the development of compounds which modulate Mitogen Activated Protein Kinases (MAPKs) for example. Therefore, an understanding the working of the components of a signal transduction opens up avenues for the future modulation of such activities with the development of new therapies and pharmaceutical agents.

The study of cell signalling can seem very daunting. Vast diagrams full of acronyms can put off the most ardent reader, but there are many basic principles which underpin the subject. In cell signalling compounds are made and initiate a response, and this is true whether the molecule originates outside the cell or is created inside. The signal transduction pathway carries a "message", with such a message originating in one place, either outside of the cell, or from another part of the same cell, but having a response elsewhere. The keys to cell signalling are that the message needs to be conveyed in a specific manner, so that it is not scrambled and misconstrued and that the cell must be able in most cases to revert back to a state or activity in which it was engaged before the message arrived, that is, the signal transduction pathway needs to be stopped when the message is no longer needed to be conveyed.

Even though the principles are simple, it is still hard to understand the complexities of cell signalling. Often signalling events are over-simplified, and components are aligned in neat rows. However, a more holistic view shows that signalling is extremely complicated and hard to understand. There are many books and chapters which explain cell signalling (Hancock, 2003; Hancock, 2010; Krauss, 2008) but these are all based on the description of the science, with the molecular details often putting off the reader who may be new to the field. Therefore, often an analogy to explain such a complex subject would be very useful, and may offer a more attractive way to teach the subject and to engage those who seek a better understanding of the area of study. In this chapter music is used as an analogy to try to shed light on some of the events in cell signalling. It has already been suggested that the use of such an analogy will be useful to those trying to get to grips with the subject (Hancock, 2005; 2009) and this chapter will expand and elaborate on those ideas. It is suggested that this can be used by those studying and teaching cell signalling.

2. Music and musical terminology

Listening to music and watching music being played are both events which rely on cell signalling. Sound waves are perceived in the ear, while photons are sensed in the eye and both lead to downstream series of signal transduction events (Hancock, 2010). In fact early work which led to the discovery of a major class of proteins, the G proteins, was due to work on the eye (Fung & Stryer, 1980). However, this is not of particular relevance to the discussion here. There is, however, growing, if controversial evidence that music can have effects on biological systems, including humans (Trappe, 2010). A term has been coined, "The Mozart Effect". This has come about from work where Mozart's music has been played and effects measured. Some tangible effects have been reported, perhaps more in the popular press than the scientific literature, but there are examples of serious reports looking into this (Jenkins, 2001).

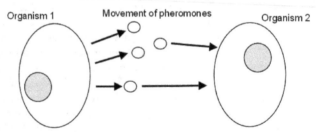

Fig. 1. Movement of pheromones transmits signals between individuals in a population. One organism releases a compound which is sensed by a second individual, in which a response is mounted. Pheromones could be thought of as music moving through the medium separating the two organisms.

Of course it is not only humans which are responsive to sound with studies reporting effects on plants for example (Qin *et al.*, 2003). Mechanical action on cells has been shown to affect cellular function (Wan *et al.*, 2004) and this includes exposure to music. For example, the activity or expression of some proteins has been shown to be changed if music is played (Chikahisa *et al.*, 2006).

Not only does music provoke cell signalling events in organisms, but music terminology is often used to describe such cellular activities. It is often said in research papers that a signalling molecule "orchestrates" or "conducts" events for example (Polo & de Fiore, 2008). In his most recent book Nick Lane uses music to explain his thinking on more than one occasion. On the theme of biological variation he discusses the musical variations of Bach and Beethoven. On the topic of protein structure he says "Yet the deeper music of the protein spheres is still there to be discerned by crystallography" and later when talking about the eye he writes "Like an orchestral conductor conjuring up the most beautiful music without sounding a note himself, the gene calls forth the structures of the eye by ushering in individual players, each with their own part to play" (Nick Lane, 2010). Therefore it is an extension of this idea which can enable music in its wider sense to act as an analogy for cell signalling. Previously the idea has been discussed (Hancock 2005; 2009) but here the ideas are expanded and enhanced.

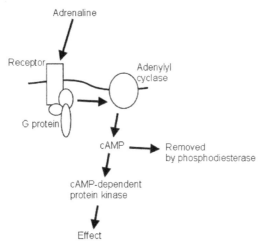

Fig. 2. cAMP is involved in the signalling invoked by adrenaline. Adrenaline is received by a plasma membrane bound receptor. The receptor is linked to a heterotrimeric G protein, which releases its alpha subunit on activation. The G protein subunit can activate adenylyl cyclase which resides in the plasma membrane. Adenylyl cyclase produces cAMP from ATP. cAMP can activate cAMP- dependent protein kinase and so lead to downstream responses. cAMP is removed by the action of phosphodiesterase.

3. Signals between organisms

Music is often produced by an individual or group of individuals, and listened to, or perceived by, another individual or group of individuals. There is an excellent example of such action in cell signalling and that is the generation and response to pheromones (Agosta, 1992; Kell et al., 1995). Here small compounds are made and released into the medium outside of the organism (Figure 1). They are then carried in the flow of the liquid or gas, perhaps the atmosphere, which surrounds the organism to be sensed and responded to by another individual of the same species, but not the individual organism that released them. Pheromones are used for attracting a mate and sexual arousal for example. In human behaviour interestingly often music is used for the same purpose.

4. Production of signals

For a cell to use a molecule in a cell signalling pathway it needs to firstly be made. Many of the components are constitutively produced and are present to partake in the required activity when called upon to do so. Examples would be large proteins such as kinases.

However, there are many situations where a molecule needs to be present, or released, in a rapid manner. Cyclic AMP (cAMP) for example is needed in response to adrenaline (for an example of a signalling pathway in which cAMP is involved see Figure 2) while insulin is released in to the blood stream when required. One of the main underlying principles of signalling is that the system is able to convey the message when and where required, in a temporal manner appropriate to the required response. Therefore molecules need to be able to partake in such signalling when called upon to do so.

There are two main ways to make a signal. Either the molecule is produced when required, or it is made and stored, to be released when required (Figure 3). It a similar manner it could be argued that there are two main ways to listen to music. You either go to a concert and in the presence of the musical instruments you listen to the sounds being made, or you let the band record the music, store it until required and then play it. In this scenario the instruments are the enzymes, producing the message. At the concert instruments make the signal as needed, to the required amount, for the required time. This is just like an enzyme such as adenylyl cyclase which is turned on, generates cyclic AMP (cAMP) for a set period of time, and then turns off. Just like the person at the concert, the protein responding to cAMP can perceive its presence and when it is all over revert back to a quiescent state – concert over.

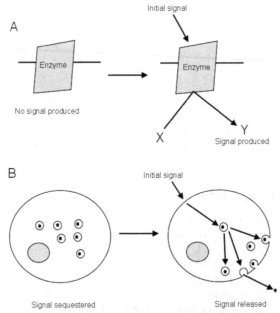

Fig. 3. Signals can be generated using different scenarios. In (A) there is no signal produced until an enzyme is activated. At that point a compound X and be converted to compound Y, so generating a signal. That is Y can now be recognised and a response or effect produced due to its presence. In (B) the signalling molecule can be pre-made but sequestered into vesicles. On arrival of the appropriate stimulus the vesicles will translocate to the plasma membrane for example. The signalling molecule will then be released and be able to move to its site of recognition and action. In the case of hormones the site of perception may be a different organ or tissue, with the signalling compound being carried by the vascular system of the organism.

The production of insulin on the other hand is more akin to a recording. Insulin is encoded for by a single gene, giving rise to a single protein, referred to as pre-proinsulin. This is heavily modified, primarily through cleave events, to produce the active insulin molecule which comprises of two polypeptides. This "ready to use" insulin is then stored in vesicles in the cell until required. This is like a musical recording, the music is created and then stored, sat in its CD case, or as an MP3 file, awaiting to be played. On demand insulin is released by the fusion of vesicles to the plasma membrane in the islets of Langerhans and is released to the outside.

5. Uniqueness of signals

Signals used by cells have to be specific and often are unique so that their presence does not get confused with another. If a signalling molecule needs to provoke a particular response it is vital that the cell's machinery recognises the presence of that specific molecule. If there was doubt then the cell may mount a response in the presence of the wrong molecule. A good example here is role of the molecules cAMP and cyclic GMP (cGMP). As can be seen in Figure 4, at first glance both these compounds look very much the same. Both have a ribose ring, a cyclic phosphate group and an added base unit.

However, the base is different in each and therefore a cell can recognise them as different. Indeed, different enzymes make them, adenylyl cyclase produces cAMP from ATP and guanylyl cyclase makes cGMP from GTP. Downstream they are recognised as separate compounds too. cAMP controls protein kinase A, while cGMP controls amongst other things protein kinase G (Figure 5).

Musical instruments are the same. Take a quick glance at a viola, and then at a violin, and they look the same. They have the same basic shape and the same basic parts. But they are different. In an orchestra they will play different music, at different times perhaps, but what is important here is just like cAMP and cGMP, a violin and a viola have their own distinct roles and parts to play in the construction of the whole, despite the fact that they are outwardly so similar.

There are occasions when cAMP and cGMP can have similar activities, and in some cases both are removed by the same phosphodiesterase. Does this mean that our analogy breaks down? Perhaps not, as if pushed different string players can pick up alternate instruments and allow the orchestra to continue. If a violin player gets ill, a viola player can often step up to the breach to fill the gap.

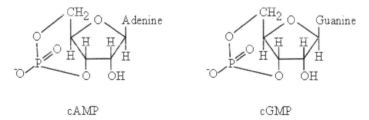

Fig. 4. Structures of signalling molecules can be very similar and yet unique. Here, the structures of cAMP and cGMP are given as examples, but they have very different signalling roles, controlling different proteins for example, that is, cAMP-dependent protein kinase and cGMP-dependent protein kinase respectively.

6. Domains and common features

In the discussion above it was argued that instruments may be similar but unique. However, there are often quite diverse instruments which share common mechanisms or structures. In signalling many proteins may also share common structures, with those structures having similar roles within the protein. A good example here is the EF hand (Lewitt-Bentley & Rety, 2000), which binds and causes a conformational change in a protein in response to changes in the levels of calcium ions in cells. EF hands can be found in a calcium controlled kinases which are able to phosphorylate downstream proteins, but EF hands are found in a wide range of other proteins too, for example the DUOX proteins involved in reactive oxygen species metabolism (Lambeth *et al.*, 2007). Although there might be subtle differences in the EF hands in different proteins the structural domain is still identifiable as being an EF hand, having the same basic function but being involved in a proteins which when taken as a whole have different functions in the cell.

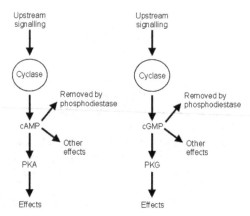

Fig. 5. cAMP and cGMP pathways are very similar. As well as cAMP and cGMP having very similar structures, if drawn in a simplistic view as shown here the signalling pathways in which they are involved is also very similar. Both are produced by cyclases, both are perceived by protein kinase, and in fact both are removed by phosphodiesterases. Both can have other effects and if viewed in more detail there are significant differences in their pathways. Importantly, cAMP and cGMP are involved in specific signalling, despite the similarities. PKA: cAMP-dependent protein kinase; PKG: cGMP-dependent protein kinase.

Therefore many signalling proteins have domains which are similar to each other and to continue our analogy musical instruments are often the same. An idea is repeated, but perhaps has a slightly different role. Consider keyboards on pianos and organs. The idea of having a set of keys which can be pressed can be found on a whole range of instruments, including different types of piano, electric keyboards, organs, harpsichords, accordions, melodicas and many others. Beyond the keyboard the mechanisms may be very different. A piano uses hammers to strike the strings, an organ opens valves to control air into pipes and so on. But the keyboard is a common feature or common structure, like the domain of a protein. And just like the EF hands discussed above, the keyboards in these different instruments may be subtly different, but they are still recognisable as being keyboards despite the overall instruments being quite different, both in shape and the type of music they play.

7. Making up a signalling system

Cells in an organism have the same genetic background, that is, they contain the same genome. The genome will encode for all the proteins which are possible to make in that organism, but different cells will have their own unique complement of proteins. Some cells will express genes for a particular signalling pathway, and others won't, perhaps having a different set of signalling proteins. Some cells will need receptors for a selection of hormones or cytokines, but other cells will have no requirement for the recognition of those extracellular signals and therefore will not express those receptors.

In music the genome is perhaps all the instruments that would be available to someone who wishes to form a group of musicians. There is a vast array of instruments available in, for example, the UK and this could be seen as the "music genome" for that country. The "music genome" for perhaps China or a Far East country may contain other instruments not commonly found in the West. So where ever a person is based to create a group of musicians they will have a "genome" to draw from, just as the cell has a genome containing the genes encoding cell signalling components to draw from. Different places could be viewed like different organisms, having different genomes, although often related. But no cell, and no music producer, would wish to enlist all the possible players. An orchestra conductor will take a wide range of instruments, from violins to kettle drums. A pop group producer may take a couple of guitars and a set of drums and little else. It is the vast array of possibilities that allow a cell to tailor its complement of proteins to allow the signalling that it needs, just as a music group will tailor its array of instruments to create the sound it requires. The cell will be able to respond to a particular group of extracellular signals while the leader of a group of musicians will be able to play a specific selection of music.

8. Receptors and their specificity

Cells are bombarded by signals all the time. It does not matter if it is a single-celled organism or a complex multicellular organism. Outside of the cell will be environmental factors such as salinity, pH, temperature, osmolarity, but on top of these will be the likely presence of compounds such as pheromones, hormones and cytokines. Therefore the cell has to be able to "decide" which it will recognise and mount a response to. The job of such a decision rests with the receptors that the cell has. If the cell synthesises the correct receptor and places it at the right place at the right time then the cell will be able to recognise and respond to the correct signal.

At the start of a practice session for an orchestra the conductor will arrive with all the scores for the piece of music which they intend to play. There will be a score for all the different instruments. The conductor generally has a score with all the parts, and s/he is like a cell which can recognise all the music. However, it would be a waste of time and effort to give such complex scores to all the musicians. Also it would make playing the music extremely unwieldy. With all the parts of the music on the score there is only a short section of notes on any page that can be seen and the musicians would be required to turn the page extremely often. This is fine for the conductor, but the musicians need to be playing, not turning pages. It would be a waste of time and effort for a cell to make all the receptors it is capable of making, that is, having the whole score. What would be the point of making a receptor for a protein hormone that it would never encounter? It would be a misuse of precious materials and space to synthesis such a receptor that will never be used. Therefore the conductor, just like an organism, ensures that each musician has the appropriate score, just with their own music to play.

This analogy can be taken further if the musicians themselves are thought of as receptors. What happens at the start of the practice session? The scores are handed out and each musician will look at the score and confirm that it is the right part, if only silently in their mind. A cello player will only wish to receive cello music: if they take violin music for example it will be in the wrong cleft and probably not able to be played. Violins not only have to get music for the violin but have the right one, being either in first or second violins. If a musician gets the wrong music they will not be able to use it and give the response that the conductor is hoping for. The musician will remain silent, even when they are supposed to be playing, like a cell receptor in the presence of the wrong ligand. It is not uncommon for musicians to return music and ask for the correct score like a receptor rejecting a ligand and leaving themselves free for the arrival of the correct ligand onto their music stand so cellular harmony can be obtained.

9. Single instruments playing a simple string of notes

Cell signalling is often presented as a neat array of components all in a line, as depicted in Figure 6A. In fact Krauss said: "The classical view of signalling pathways has been that of sequential transmission of signals in a linear signalling chain". This would be like listening to solo instruments, playing a linear line of notes on a page. Some representations of cell signalling show little else, making the systems look simple and easy to understand. But just as we don't often listen to single instruments, unless listening to a sonata perhaps, cell signalling is also a complex mix of many players, all adding to the harmony at the same time. As Krauss goes on to point out, signalling is far more complicated (Krauss, 2008).

Fig. 6. Signalling pathways are often depicted as a single series of components in a line, akin to a single line of music (A). However, they are usually much more complex, more like several instruments all playing at the same time (B).

Therefore cell signalling should be thought of as a group of instruments and voices, all competing for attention at the same time. Perhaps it is more like depicted in Figure 6B. Just as in a musical group, not all these musicians need to be in action all the time, and in fact it could be rather boring if that was the case. But they will all be there on stage, awaiting their cue from the music, always ready for action, but only acting when needed. In a cell the situation is the same, many signalling components will be quiescent until they are drawn upon to play their role in the control of the cell.

10. Degeneracy

It is often a puzzle in molecular biology that proteins may be able to replace each other. Perhaps an inhibitor has been added which is supposed to remove the functioning of a cell signalling component, but the effect is far less than anticipated. In knock-out or knock-down studies, where the expression of a protein is completed ablated, or severely reduced, then the cell sometimes shows little effect (see Colucci-Guyon *et al.*, 1994 as an example). In a cell signalling response, again often far less than anticipated is seen than from the theory (see Zhang *et al.*, 1994 as an example). In such situations the most likely scenario is that proteins are replacing each other when needed, and protein function is said to be degenerate.

If proteins can replace each other in function can we again invoke a music analogy. The leader of the orchestra is a very important position, often helping and advising the conductor during practise sessions, but during the concert they are not redundant when it comes to control either. Like a signalling component in cells, the leader will be signalling to the rest of the violins when to start playing, so that the whole section plays together and sounds like one. However, what happens when the leader breaks a string. Violin strings are under a large amount of tension and break quite often. If the leader's violin breaks, does the orchestra have to stop. In reality, especially in a rehearsal, it probably would and a new string would be rapidly fitted. But if as in a cell, waiting for a response is not a pragmatic option and the orchestra has to continue then another violin player, probably the one in the second seat next to the leader, will take over and allow the orchestra to continue. One musician, just like the cell's protein, will take up the important role vacated by the other.

Other scenarios in music also can be envisaged. Often musicians can play more than one instrument, and especially in non-professional orchestras and groups it would not be uncommon for one player to take the place of an absence colleague. This would explain why the absence of what was thought to be an important player, or protein, can be seen to not have a devastating effect on the harmony of the music group, or cell.

11. Multiple functions of some proteins

It has been recognised that proteins often have more than one role and are said to be "moonlighting" (Jeffrey, 2009). Here, as well as proteins being able to cover the roles of each other as in degeneracy, some proteins have other very disparate roles.

Some proteins had roles discerned many years ago only to have new additional roles assigned to them more recently. A good example here is cytochrome c which was for many years assigned to a redox role in the mitochondria, only later to be found to be instrumental in controlling apoptosis (Figure 7. Reviewed by Jiang & Wang, 2004). Glyceraldehyde 3-phosphate dehydrogenase is central to glycolysis, but it too has now been assigned cell signalling roles, in particular as a protein which translocates to the nucleus to control gene expression (Tristan *et al.*, 2011).

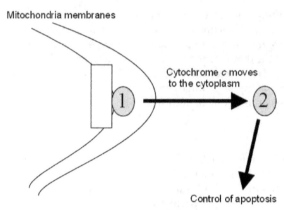

Mitochondria membranes

Cytochrome *c* moves
to the cytoplasm

Control of apoptosis

Fig. 7. Cytochrome *c* has more than one function. Cytochrome *c* is normally found in the mitochondria. It resides associated with the inner mitochondrial membrane, where it acts to shuttle electrons from Complex III to Complex IV. However, there is a signalling pathway which leads to cell suicide, or apoptosis, in which cytochrome *c* leaves the mitochondria and moves into the cytoplasm. Here it interacts with the caspase system which leads to ultimately to cell death. Therefore cytochrome *c* has two very distinct and disparate functions.

Musicians and their instruments are like such moonlighting proteins. They may play a violin, but in fact such instruments can be used in multiple music genres. During the day the violinist may be in an orchestra, but some evenings may be playing jazz in the local bar. As with the proteins, their roles are not fixed, and translocation from one venue to another will allow them to partake in a new role. Furthermore, many if not most musicians are experts at more than one instruments. Therefore if an orchestra is short of a viola player, then perhaps a violinist can take their place. Or the change of instrument can be more dramatic, with a flutist can take over on the kettle drums. Just like many proteins, temporal and spatial location of the player, and the interactions in which they may partake, may dictate the exact role they play at any moment in time.

12. Subtle changes make a big difference

In signalling systems there are many components all vying for attention, and to initiate a response. However, even though it is individual components which are studied in many cases, the overall response of the cell will be dictated by the sum of the signalling which is taking place, an idea modelled by Rachmilewitz & Lanzavecchia (2002). Even so, within this holistic approach it needs to be realised that subtle changes in the levels of some signalling components can give a large effect even if there is no change in other proteins and molecules. Certainly some signals have been described as dominant over others. An example here would be a paper by Reya and Clevers (2005) who write "Current evidence indicates that the Wnt cascade is the single most dominant force in controlling cell fate along the crypt–villus axis". Therefore, a small change in a dominant pathway would initiate a significant response regardless of the activity of less dominant pathways.

Music is often like this, with the single notes in amongst many others having a profound effect. If a pianist plays a major chord, perhaps a C, E and G, to produce a chord of C major, the effect is recognisable, and being major will sound cheerful. However, keeping the C and G the same, but lowering the tone of the E by a semitone to E flat and the chord becomes minor. This is a significant effect. One note amongst the three has changed by a relatively small amount, but the chord is now recognisably different, and the resultant sound has gone from a happy major to a rather sad minor. A small musical change, with a large result. Beethoven uses the alteration of one note in amongst many to great effect in the first movement of the "*Moonlight Sonata*" (Piano Sonata no. 14 in C sharp minor - Op. 27 no 2) for example.

13. Background, volume and thresholds

Signalling in cells needs to take place in the presence a background level of "signalling noise". Cells are bombarded by extracellular signals all the time, whether from the environment of the organism or from other cells of the same organism. The demands on a cell will be constant and varied in many cases. Therefore, if a major response is needed, the signalling that is invoked needs to be "heard" above the noise of the rest of the activity of the cell.

Most signalling systems will in fact be in a state of equilibrium. Often levels of signalling components are measured, perhaps before and after a treatment. However, rarely do the levels of activity, levels of signalling molecules or levels of phosphorylation go from a base level of zero to a higher level. In the vast majority of cases the levels rise from a low level to a transiently higher level. Therefore researchers define threshold levels for signals, (for example Pereyra *et al.*, 2000).

How individual signals get heard may be liken to phrases in the orchestra when one instrument temporally is dominate and can be clearly heard, especially an instrument such as a kettle drum. In music there are often many instruments play all at the same time, and often it is hard to discern the exact contribution of any one instrument. The holistic effect may be pleasing but the parts played by individuals are assumed to be part of the whole. However, one instrument can dominate over the others and be heard above the rest. Perhaps a trumpet is playing a strident part. It will be heard above the other hundred instruments which make up the orchestra. And the audience will follow the trumpet, the tune from which will carry the music and the mood. Cells will have a similar system. Many signals are all contributing, but the arrival of a new hormone may need to dominate. A pathway may be activated, and the activity of the players in that pathway will reach a threshold allowing them to have their effect above that of the other signalling components, which will after all be carrying on doing what they were doing before. However, transiently, the pathway with the "volume" which is dominant will be able to invoke the cellular response needed.

Some instruments such as a bagpipe rely on a background tone, or drone (Nordquist & Ayers, 2009). The highland bagpipe has a tonic note (that is the base note of the scale) of A. Therefore, the other notes are played over this, but the tonic gives the constant tone to the music. It is the other notes which will dominate to give the tune and harmonies. In cells there are various parameters upon which activities and functions of proteins and signalling components will need to contend. In cellular compartments pH is crucial, but as well as this there is the redox state of the environment. In the cytoplasm for example, there is a high concentration of reduced glutathione which will endeavour to maintain the redox state relatively constant (Schafer & Buettner, 2001). This is important because proteins contain reactive side groups which may be affected by the redox of the medium in which they function. Such groups include the thiol groups of cysteine residues for instance. Here, two cysteines may react together in an oxidation reaction to create a disulphide bridge which may stabilise the protein. Alternatively they may react with signalling molecules such as nitric oxide (to be S-nitrosylated) or hydrogen peroxide (to be oxidised). Disruption of the redox state of the cell towards the oxidised state is referred to as oxidative stress, a condition of cells with is extremely important not only to control cell function but also to regulate processes such as apoptosis. Oxidative stress has been implicated in numerous diseases, including degenerative disease (Kadenbach et al., 2009). Therefore it is extremely important for this basal redox state to be maintained, much like a basal note of the bagpipe. It needs to be there, allowing continuity of the harmony of the cell. However, there does need to be the involvement of signals such as hydrogen peroxide and nitric oxide. It may be that the basal redox state in some cases maintains the thiol groups in a state to enable compounds such as nitric oxide to react and have its effect. This would be like the pitched notes on the bagpipe being strident above the background tonal level. On the other hand, if the background is disturbed, perhaps during oxidative stress, such thiols would have already reacted with for example hydrogen peroxide and be no longer available for a reaction it would normally partake in. A disruption of the basal background harmony has altered the effect of the other signals, and the overall effect is quite different. For normal signalling to resume, the background "tonal" redox state would need to be restored.

14. Timing and phasing: Oscillations and waves

One of the intriguing aspects of cell signalling is the timing of the signals and how they fit together temporally. To get a full understanding of signalling needs a full appreciation of both the spatial and temporal aspects of any signal, but particularly how they might be working together in time and space. Early work in this area concentrated on calcium ion signalling, and it was reported that calcium ions were not only altered transiently in some systems but this transient change in ion concentrations actually followed an oscillating pattern. A superb example of this is shown by Alberts et al. (1994). Here, the oscillations are dependant on the concentration of the initial signal added. It is not the amplitude of the change which seems to be important in this signalling, but rather the frequency of the oscillations. However, temporal fluctuations on the concentrations of signals are not unique to calcium ions. The biphasic nature of other signalling systems has also been reported, for example with reactive oxygen species (Bleeke et al., 2004) and also with insulin signalling (Rorsman et al., 2000). If hydrogen peroxide levels are followed for example, they increase quickly but transiently, but after a period of relatively low activity the levels once again rise, often to be sustained for the second period. This may be reflected in levels of other signals too, such as nitric oxide. Therefore, at any moment in time the levels of signals may be rising and falling, and it is probably the combined nature of such changes which brings about the desired response in the cell. It is pattern of change which should be considered, rather than the individual changes which might be being recorded.

Music is often written in a pattern. As discussed above, Lane likens biological variations and patterns to musical variations (Lane, 2010). But musical patterns are often phased too. A prime example here is the fugue. Oxford Dictionaries describe a fugue as being written in such as way that "...a short melody or phrase (the subject) is introduced by one part and successively taken up by others and developed by interweaving the parts." A superlative example of such a work is the fugue in the *Toccata and Fugue in D minor*, BWV 565, by Johann Sebastian Bach.

Cell signalling in some cases needs to be thought of in this manner. Hydrogen peroxide and nitric oxide can be considered as two lines of music, one being interwoven with the other. One rising and falling in unison with the changes in the other. But of course it would be naïve to think in these terms for just two signals such as these. Both nitric oxide and hydrogen peroxide can impinge on calcium signalling, so phasing of changes in calcium ion concentrations will need to be considered too. But a myriad of other signals will be employed at any moment in time in a cell so the overall response, or set of responses needs to be orchestrated by the phasing and overall shifting pattern of signals being employed by the cell.

15. Setting up for the future

In music there are often times when the phrasing and harmony just does not sound quite right. This is usually very transient and the harmonies resolve very quickly. Perhaps a composer has asked for a F and a G to be played together. If they were the adjacent notes the resulting discord would be very obvious, but often such notes are played with two or three octaves between them – in those cases the discord is not so blatant. However, often that harsh nature of the discord will mean that when the music does harmonise the end result is more pleasing than it would have been without the disharmony. The composer has set the scene for the final resolution. Again using Beethoven as an example, he does this to great effect in the "*Moonlight Sonata*", where the listener is treated to slight disharmony and one is waiting in anticipation for the resolution, which when it comes is delightful. It brings depth and feeling to the work. Therefore the composer is setting up for the future, ensuring that what subsequently arrives results in a success.

Cells need constantly to be setting the scene and making sure that they are ready for the future. And of course cell signalling is the key to doing this. Signalling often leads to adaptation, where the cell sets itself up for future possible events (Neill *et al.*, 2002). Music often sets the scene in a strident and discordant, or stressful, way to allow for future harmony. In cells exposure to one stress can lead to cells being able to cope better with subsequent stress in the future, and not only to the same stress. Temperature stress in plants for example can lead to adaption to future stress by other abiotic and biotic stress factors. Instead of viewing sub-lethal stress as a negative thing in cell signalling perhaps we should be more like the composer who is prepared to chose a discord to ensure the future has a better outcome. The composer is adapting our ear, just as cell signalling is adapting the functioning of the cell for future events.

In cell signalling adaptation and preparing for the future may require long term activity, and will no doubt involve the control of gene expression with an alteration of the complement of proteins in the cell. Perhaps the cell will alter its levels of certain receptors or signalling proteins, and this will enable the cell to have a faster or more tailored response in the future. This would be like a disc jockey in the night club being asked for a certain genre of music, but realising that he didn't have it. There would be a period of short term stress. Instead of being caught out in the future, a trip to the music shop would ensure that he would be "adapted" for future requests, so lessening the chance of stress, and also allowing alternatives to be played which may relieve an otherwise stressful situation. Our cell signalling would "take a trip to our genome" and so ensure that they are ready for future "requests" from their environment.

16. Signalling dysfunction

Dysfunction of cell signalling can have catastrophic consequences. This is certainly one of the main reasons why the topic needs to be more fully understood. Dysfunction can lead to either the lack of functionality or indeed too much activity, with either situation being undesirable. If insulin signalling is taken as an example, a dysfunction of the insulin receptor would mean that the arrival of insulin at the cell surface would not be recognised and no insulin response by the cell would be mounted. Obvious effects of such a dysfunction are conditions such as diabetes. One the other hand, if the G protein Ras is taken as an example, mutation of the coding for amino acids at position 12, 13, or 61 in the sequence leads to a protein which has impaired GTPase activity and therefore can not be turned off (Figure 8). This leads to the signalling pathways in which RAS is involved being in the permanently active state, regardless of the lack of continued initiation of the signalling pathway. *RAS* mutations are found in about one third of human malignancies (Riely *et al.*, 2009). These are just two examples and there are many more, highlighting the importance and impact of correct cell signalling.

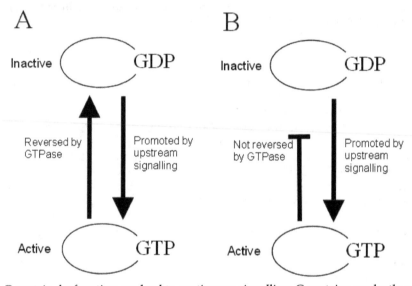

Fig. 8. G protein dysfunction can lead to continuous signalling. G proteins can be thought of as being molecular switches, with an "on" and an "off" state. In the inactive state they are bound to GDP, but on activation this is exchanged for GTP, so leading to a conformational change in the protein which allows it to signal (A). The GTP bound form of the G protein will then signal downstream to the next effector in the chain. In the case of the G protein Ras, the next signalling component in the transduction pathway could be the protein kinase Raf. To then inactivate the G protein its intrinsic GTPase activity will convert the GTP back to GDP and inactive the protein – through a reversal of the conformation change. However, in proteins such as Ras, a mutation can disrupt the GTPase active site so stopping the conversion of GTP back to GDP (B). In this case, the protein will continue to be bound to GTP, and continue to be in the conformation that signals. Therefore, even if all the upstream signalling is reversed or halted, the G protein will continue to signal downstream regardless. Because Ras is often on pathways which are invoked by growth factors, continuous G protein signalling can lead to a continuous "grow" signal, and hence lead to tumour growth and cancer.

Dysfunction can and does happen in music too. The discussion above emphasises the fact that many cell signalling events will be taking place in the cell at the same time. There may be many effects, often in different parts of the cell. However, there should be minimal interference of one pathway over the other if they are controlling completely independent effects. At many music venues in recent times there may be several events taking place all at the same time, and the Glastonbury Festival is a good example. People need to listen to the band of their choice without hearing the others in such a way that it disrupts their enjoyment. However, if one band starts to dominate, or the equipment on one band loses its volume, then the effect that the festival envisaged will be compromised, just as the overall signalling network of the cell would be compromised. An example of this would be signals moving into a cell through the gap junctions. The signalling of one cell may overwhelm the signalling in one of its neighbours if second messengers move on mass through the gap junctions. Clearly there needs to be control of such movement and gap junction function and regulation is clearly important to understand (Evans and Martin, 2002).

Although it is less common now physical recordings can cause problems with both vinyl records and CDs able to "jump". This can render the music so bad that it can't be listened too. Other equipment can fail too, including digital instruments and amplifiers. But even more classical equipment can have problems. A sonata played on a piano with a broken key or hammer may make the music very poor. It may not stop the piece being played altogether, and the musician may be able to continue. However, the concert is unlikely to get good reviews and both the musician and venue may struggle to have a future event. Just like in a cell, a small dysfunction may render the longer term future to be in doubt.

17. Future and evolution

It would be naïve to think that cell signalling has evolved to the point where it will evolve no further. Organisms continue to evolve and the proteins involved in cell signalling will no doubt evolve too, and certainly will not stay the same for eternity. Likewise it would be foolish to think the same about music and instruments.

Over billions of years since life began, cell signalling proteins have mutated and changed to give the polypeptides that we can find today. Since the creation of the first cells some form of signalling was required, both to sense the environment of the cell and to coordinate adaption as the world changed. From an oxygen free atmosphere to the present climate of Earth cells and their signalling have had to adapt along the way. They will continue to change, adapt and no doubt the proteins involved in signalling will increase in number in the future. Perhaps an example of such a change can be seen with the enzyme nitric oxide synthase (NOS). Very recently a NOS has been characterised from a very primitive green algae *Ostreococcus tauri* (Foresi *et al.*, 2010). Perhaps this photosynthetic organism inherited its gene for this enzyme from a more primitive cell, one which gave rise to both plants and animals. This is likely as the *O. tauri* amino acid sequence is 45% similar to that of a human gene for NOS. Therefore, the human gene has changed considerably compared to the *O. tauri* gene over time. Perhaps more striking is the fact that although *O. tauri* is a primitive plant, higher plants do not seem to have a form of this NOS gene at all. Either it has been lost altogether, or it has been mutated to a form which has yet to be identified. Either way, evolution has been hard at work on this gene, and will no doubt continue such work into the future.

Mutation and duplication of gene can lead to families of proteins. Certainly in signalling families of protein isoforms can be recognised, some with added domains, some with extra phosphorylation sites and some in truncated form. Good examples are phosphatase proteins that remove the phosphate groups from proteins (Cohen *et al.*, 1990).

Musical instruments also evolve, and in many ways in a similar manner to proteins. Some have certainly been around for a long time, but even those that form part of current orchestrates are different from those used by great composers such as Mozart. Furthermore he would never have imagined the possibility of an electric violin, but today his music is often played on such an instrument. Music itself evolves, with successive composers building on the work of those who went before them.

Instruments have changed over the years in a way that resembles that of proteins. Protein isoforms can be created when a gene is copied so there are two versions, and then those genes mutate after a period of time to two separate genes which are able to be characterised, and they would give rise to different but related proteins. Musical instruments are the same. A violin is like a copy of a viola, except one is bigger and plays different notes. Copy it again and make it bigger still and a cello is created, and so on so there is a family of instruments which are recognisable as being related, and yet they have different roles. They could be thought of as isoforms perhaps, just like proteins. Using the piano as an example and again one can see "isoforms" which are all recognisable as pianos, that is the concert grand, the baby grand, the upright, the studio piano and so on. Over the years the piano has been adapted to the place it needs to be placed and the audience it is aimed at. There have formats which are no longer seen, like genes which have disappeared during evolution, and there are new versions being developed and used. There now seems to be a vast array of electric pianos and electronic keyboards, with the idea of using a piano keyboard layout being copied and mutated to develop new instruments.

Music and the instruments used to create it will continue to develop and evolve, just as the proteins which are involved in cell signalling. Especially in the advance of climate change, organisms will need to adapt and it will be cell signalling which coordinates such changes, but the signalling pathways will change too. The future will see the development of new cellular components, and no doubt new musical instruments, especially as new digital technologies are adopted. Not all changes will be beneficial, with mutations in genes no only allowing the future evolution of species but creating dysfunctional proteins along the way causing disease in individuals. No doubt not all new music innovations will be successful either, and the future will be littered with new proteins and musical instruments abandoned by nature and the music industry respectively.

18. Conclusion

Cell signalling is both enormously important to the understanding of how cells works and immensely complicated. Therefore ideas which can be used to aid in the teaching and study of the subject would be extremely helpful, and an analogy is often a good tool. Music initiates cell signalling events in organisms, but music terminologies are often used to explain aspects of cell signalling. However, music as an analogy for cell signalling events can be an interesting and useful way to look at the principles of signalling and transduction pathways. Such an analogy will be useful to those teaching and studying cell signalling.

19. Acknowledgements

I would like to thank Annabel Hancock for supplying the music used in Figure 6.

20. References

Agosta, W.C. (1992) *Chemical communications: the language of pheromones.* Scientific American Library, New York. USA,.

Alberts, B., Bray, D., Lewis, J., Raff, M., Roberts, K. & Watson, J.D. (1994) *Molecular Biology of the Cell* 3rd edn. Garland Press, New York.

Bleeke, T., Zhang, H., Madamanchi, N., Patterson, C. & Faber, J.E. (2004) Catecholamine-induced vascular wall growth is dependent on generation of reactive oxygen species. *Circ. Res.,* Vol. 94, pp. 37–45.

Chikahisa, S., Sei, H., Morishma, M., Sano, A., Kitaoka, K., Nakaya, Y. & Morita, Y. (2006) Exposure to music in the perinatal period enhances learning performance and alters BDNF/TrkB signalling in mice as adults. *Behav. Brain Res.,* Vol. 169, pp. 312–9.

Cohen, P.T.W., Brewis, N.D., Hughes, V. & Mann, D.J. (1990) Protein serine/threonine phosphatases: an expanding family. *FEBS Lett.,* Vol. 268, pp. 355–359.

Colucci-Guyon, E., Portier, M.M., Dunia, I., Paulin, D., Pournin, S. & Babinet, C. (1994) Mice lacking vimentin develop and reproduce without an obvious phenotype. *Cell,* Vol. 79, pp.679-694.

Evans, W.H. & Martin, P.E.M. (2002) Gap junctions: structure and function (Review). *Molecular Membrane Biology,* Vol. 19, No. 2, pp. 121-136.

Dawkins, R. *The Selfish Gene* (1989) 2nd edn. Oxford University Press, Oxford, UK

Filmore, D. (2004) It's a GPCR world, cell-based screening assays and structural studies are fueling G-protein-coupled receptors as one of the most important classes of investigational drug targets. *Modern Drug Discovery,* Vol. 7, pp. 24–28.

Foresi, N., Correa-Aragunde, N., Parisi, G., Caló, G., Salerno G. & Lamattina L. (2010) Characterisation of a nitric oxide synthase from the plant kingdom: NO generation from the green algae *Ostreococcus tauri* is light irradiance and growth phase dependent. *The Plant Cell,* Vol. 22, pp. 3816-3830.

Fung, B. K.-K. & Stryer, L. (1980) Photolyzed rhodopsin catalyses the exchange of GTP for bound GDP in retinal rod outer segments. *Proc. Nat. Acad. Sci. USA,* Vol. 77, pp. 2500-2504.

Jenkins, J.S. (2001) The Mozart effect. *J. R. Soc. Med.,* Vol. 94, pp. 170-172.

Hancock, J.T. (2009) Cell signalling is the music of life. *Brit. J. Biomed. Sci.,* Vol. 65, pp. 205-208.

Hancock, J.T. (2010) *Cell Signalling.* 3rd edn. Oxford Univeristy Press, Oxford, UK.

Hancock, J.T. (2005) *Cell signalling.* 2nd edn. Oxford University Press, Oxford UK.

Hancock, J.T. (2003) Principles of cell signalling. In *On growth, form and computers.* Kumar, S., Bentley, P.J. eds. pp. 64–81, Academic Press, Oxford.

Jeffrey, C.J. (2009) Moonlighting proteins—an update. *Mol. BioSyst.,* Vol. 5, pp. 345-350.

Jiang, X. & Wang, X. (2004) Cytochrome C-mediated apoptosis. *Annu. Rev. Biochem.,* Vol. 73, pp. 87-106.

Kadenbach, B., Ramzan, R. & Vogt, S. (2009) Degenerative diseases, oxidative stress and cytochrome c oxidase function. *Trends in Molecular Medicine,* Vol. 15, No.4, pp. 139-147.

Kell, D.B., Kaprelyants, A.S. & Grafen, A. (1995) Pheromones, social behaviour and the functions of secondary metabolism in bacteria. *Trends Ecol. Evol.* Vol. 10, pp. 126–9.

Krauss, G. (2008) *Biochemistry of Signal Transduction and Regulation*. Wiley-VCH, Chichester, UK.

Lambeth, J.D., Kawahara, T. & Diebold, B. (2007) Regulation of Nox and Duox enzymatic activity and expression. *Free Radic. Biol. Med.*, Vol. 43, pp. 319-331.

Lane, N. (2010) *Life Ascending: The Ten Great Inventions of Evolution*. Profile Books Ltd, London, UK.

Lewitt-Bentley, A. & Rety, S. (2000) EF-hand calcium-binding proteins. *Current Opinion in Structural Biology*, Vol. 10, pp. 637-643.

Neill, S.J., Desikan, R., Clarke, A., Hurst, R. & Hancock JT. (2002) Hydrogen peroxide and nitric oxide as signalling molecules in plants. *J. Exp. Bot.*, Vol. 53, pp. 1237-1247.

Nordquist, P.R. & Ayers, R.D. (2009) Tuning and tone quality of bagpipe drones. *Acoust. Soc. Am.*, Vol. 125, pp. 2652-2652.

Oxford Dictionaries. http://oxforddictionaries.com/

Pereyra, E., Mizyrycki, C. & Moreno, S. (2000) Threshold level of protein kinase A activity and polarized growth in *Mucor rouxii*. *Microbiology*, Vol. 146, pp. 1949-1958.

Polo, S. & de Fiore, P.P. (2008) Endocytosis conducts the cell signalling orchestra. *Cell*, Vol. 124, pp. 897–900.

Qin, Y.C., Lee, W.C., Choi, Y.C. & Kim, T.W. (2003) Biochemical and physiological changes in plants as a result of different sonic exposures. *Ultrasonics*, Vol. 41, pp. 407-411.

Rachmilewitz, J. & Lanzavecchia, A. (2002) A temporal and spatial summation model for T-cell activation: signal integration and antigen decoding. *Trends Immunol.*, Vol 23, pp. 592-595.

Riely, G.J., Marks, J. & Pao, W. (2009) *KRAS* mutations in Non–Small Cell Lung cancer. *The Proceedings of the American Thoracic Society*, Vol. 6, pp. 201-205.

Reya, T. & Clevers, H. (2005) Wnt signalling in stem cells and cancer. *Nature*, Vol. 434, pp. 843-850.

Rorsman, P., Eliasson, L., Renström, E., Gromada, J., Barg, S. & Göpel, S. (2000) The cell physiology of biphasic insulin secretion. *News Physiol. Sci.*, Vol 15, pp. 72–77.

Schafer, F.Q. & Buettner, G.R. (2001) Redox environment of the cell as viewed through the redox state of the glutathione disulfide/glutathione couple. *Free Radic. Biol. Med.* Vol. 30, No. 11, pp. 1191-1212.

Trappe, H.J. (2010) The effects of music on the cardiovascular system and cardiovascular health, *Heart*, Vol. 96, pp. 1868-1871.

Tristan, C., Shahani, N., Sedlak, T.W. & Sawa, A. (2011) The diverse functions of GAPDH: views from different subcellular compartments. *Cell Signal.* Vol. 23, pp. 317-323.

Wan, X., Steudle, E. & Hartung, W. (2004) Gating of water channels (aquaporins) in cortical cells of young corn roots by mechanical stimuli (pressure pulses): effects of ABA and of $HgCl_2$. *J. Exp. Bot.* Vol. 55, pp. 411–422.

Zhang, R., Tsai, F.Y. & Orkin, S.H. (1994) Hematopoietic development of vav-/- mouse embryonic stem cells. *Proc. Nat. Acad. Sci. USA*, Vol. 91, pp. 12755-12759.

4

AGE/RAGE as a Mediator of Insulin Resistance or Metabolic Syndrome: Another Aspect of Metabolic Memory?

Hidenori Koyama and Tetsuya Yamamoto

Department of Internal Medicine, Division of Endocrinology and Metabolism,
Hyogo College of Medicine,
Japan

1. Introduction

Large randomized studies in diabetes have established that early intensive glycemic control reduces the risk of diabetic microvascular complications, with less impact on macrovascular complications [1, 2]. In type 2 diabetic patients, further intensive therapy to target normal glycated hemoglobin levels also failed to reduce mortality and major cardiovascular events [3, 4], while it may be rather harmful [5]. However, follow-up data of these trials reveal a long-term influence of early metabolic control on longer cardiovascular outcomes, even though the influence on glycemic control has been immediately disappeared after the trials [6, 7]. This phenomenon has recently been defined as "metabolic memory". In at-risk patients with type 2 diabetes, intensive intervention with multiple drug combinations and behavior modification had similar sustained beneficial effects with respect to vascular complications and on rates of death from any cause and from cardiovascular causes [8]. Similarly in patients with end-stage renal disease (ESRD), intensive interventions to the general risk factors, such as high LDL-cholesterol or C-reactive protein, have not been successful in improving their cardiovascular outcomes [9, 10], suggesting that the beneficial effect of risk reduction may be overwhelmed by accumulated "metabolic memory" by long-term exposure to oxidative stress during the progression of renal failure.

Potential mechanisms for propagating this "memory" are the non-enzymatic glycation of cellular and tissue proteins which are conceptualized as advanced glycation end-products (AGEs), the generation of which has been implicated to be deeply associated with increased oxidative stress as well as hyperglycemia. AGEs, with their receptor (receptor for AGEs, RAGE), potentially mediate molecular and cellular pathway leading to metabolic memory. Moreover, interaction of the RAGE with AGEs leads to crucial biomedical pathway generating intracellular oxidative stress and inflammatory mediators, which could result in further amplification of the pathway involved in AGE generation.

By utilizing genetically engineered mouse models, emerging evidence suggests that AGE/RAGE axis is also found to be profoundly associated with non-diabetic, non-uremic pathophysiological conditions including 1) atherogenesis, 2) angiogenic response, 3) vascular injury, and 4) inflammatory response (see review in [11]), many of which are now implicated in metabolic syndrome. Numerous truncated forms of RAGE have also been described, and the

C-terminally truncated soluble form of RAGE has received much attention. Soluble RAGE consists of several forms including endogenous secretory RAGE (esRAGE) which is a spliced variant of RAGE [12], and a shedded form derived from cell surface RAGE [13, 14]. These heterogeneous forms of soluble RAGE, carrying all of the extracellular domains but devoid of the transmembrane and intracytoplasmic domains, bind ligands including AGEs, and may antagonize RAGE signaling in vitro and in vivo. ELISA systems to measure plasma esRAGE and total soluble RAGE have been developed, and decreased plasma esRAGE is found to be associated with insulin resistance, obesity and metabolic syndrome [15]. Moreover, our recent observation highlights the direct role of RAGE in adiposity; RAGE deficiency is associated with less weight gain, less abdominal fat mass, less adipocyte size, less atherosclerotic lesion formation and higher plasma adiponectin than wild type control [16].

Insulin resistance is the primary mechanism underlying the development of type 2 diabetes and is a central component defining the metabolic syndrome, a constellation of abnormalities including obesity, hypertension, glucose intolerance, and dyslipidemia. Insulin resistance or metabolic syndrome has been defined to be associated with low-grade inflammation, and therefore inflammation could contribute in large part to its development [17], implicating an intriguing possibility that this pathophysiological condition is also an additional face of metabolic memory driven by RAGE axis. Although insulin resistance has been characterized by complex factors including genetic determinants, nutritional factors, and lifestyle, growing evidence suggests that mediators synthesized from inflammatory cells are critically involved in the regulation of insulin action. In brief, insulin binding to its specific receptor stimulates tyrosine phosphorylation of insulin receptor substrate (IRS) proteins, which is a crucial step for insulin signaling system. Many inflammatory signals appear to induce serine phosphorylation of IRS, which could be involved in disruption of insulin-receptor signaling [17]. In this chapter, we would like to summarize the recent findings regarding pathophysiological roles of RAGE and soluble RAGE in insulin resistance and metabolic syndrome.

2. AGEs

2.1 AGE formation by glucose and its derivatives

AGEs are proteins generated by a series of reactions termed the Maillard Reaction. Classically, AGE formation has been described by a nonenzymatic reaction between proteins and glucose [18, 19]. AGEs derive from the spontaneous reaction of carbohydrates with amino group of proteins, which undergo from the formation of reversible products (Schiff base adducts) to the generation of more stable products (Amadori products). Subsequently, complex reactions occur including intermolecular crosslink formation, and cleavage through oxidation, dehydration, condensation, cyclization, and other reactions follows, with generation of AGEs through a late reaction characterized by fluorescent and brown coloration and molecular crosslinkage. Recently, it was confirmed that AGEs are also formed by non-enzymatic reaction of reactive carbonyl compounds such as 3-deoxyglucosone, methylglyoxal resulting from persisting high blood glucose level, and oxidative stress associated with the amino residues of proteins.

2.2 AGEs formation independent of hyperglycemia

There is also increasing evidence that AGEs are also formed through lipid-derived intermediates, resulting in advanced lipoxidation products [20]. AGEs might be formed directly by autoxidation of free glucose [21, 22]. In this pathway, known as autoxidative glycosylation, such reactive oxygen species as hydrogen peroxide were identified as both products and catalysts of autoxidation of sugars. Other than diabetes mellitus patients, high plasma and tissue levels of AGEs are observed in patients with ESRD. It has been reported that no difference was noted in blood AGEs levels between those with and without diabetes mellitus among chronic renal failure patients on hemodialysis, which is believed to enhance production and accumulation of AGEs in conditions other than hyperglycemia. Local accumulation of AGEs is also observed in patients with Alzheimer disease, rheumatoid arthritis, arteriosclerosis, cancer, and other diseases, suggesting the involvement of inflammation and oxidative stress in the formation of AGEs.

2.3 Orally absorbed AGEs

In addition to the endogenously formed, AGEs are abundant in exogenous sources such as foods, especially when prepared under elevated temperatures [23]. Vlassara's group has extensively examined the role of the exogenous AGEs in several pathological conditions (see review in [24]). After ingestion, 10% of orally-administered AGEs are absorbed into the circulation [25-27], majorities of which are shown to be accumulated in tissues. Among them are tissue-reactive α, β-dicarbonyl-containing intermediate products, such as methylglyoxal, which has been linked to cellular oxidant stress and apoptosis [28], and terminal products, such as $^\varepsilon$ N-carboxymethyllysine (CML), which is formed by glycoxidation as well as by lipoxidation [29-31]. Both methylglyoxal and CML have been identified in vivo and are shown to be associated with oxidant stress and tissue damage [31-33].

3. AGEs and endogenous RAGE ligand in insulin resistance or metabolic syndrome

3.1 AGEs in insulin producing and acting tissues

Importantly, among the multiple targets of bioactive AGEs are also such diverse tissues as the pancreatic islet [34, 35], the adipose tissue (adipocyte) [36] and skeletal muscle cells [37], major tissues involved in insulin secretion and its actions. Pharmacological inhibition of glycoxidation protects against damage to either tissue [33, 38]. AGEs are shown to inhibit glucose-stimulated insulin secretion from islet through iNOS-dependent nitric oxide production [35]. Reduced intake of dietary AGEs has also been shown to decrease the incidence of type 1 diabetes in NOD mice [39] as well as the formation of atherosclerotic lesions in diabetic apolipoprotein E-deficient mice [40].

3.2 AGEs and insulin resistance in vivo

AGEs burden is also shown to be associated with impaired endothelial function [41-44]. Endothelial dysfunction could be profoundly associated with less insulin delivery to the skeletal muscle interstitium, leading to decreased insulin-stimulated glucose uptake by the skeletal muscle [45-49], which is implicated in the pathogenesis of insulin resistance. More directly, the restriction of the AGE content in standard mouse diets was found to markedly improve insulin resistance in obese $db/db^{(++)}$ mice [50]. More recent observation shows targeted reduction of the advanced glycation pathway improved renal function in obesity [51]. This interesting observation was further supported by the findings that the development of insulin resistance and type 2 diabetes during prolonged high-fat feeding are linked to the excess AGEs/advanced lipoxidation end products inherent in fatty diets [52].

3.3 AGEs and insulin resistance in vitro

Several evidences also suggest that AGEs affect the function of insulin-target cells in vitro. AGEs interact with CD36 in mouse 3T3 and human subcutaneous adipocytes, which is associated with down-regulation of leptin expression in adipocyte through reactive oxygen species (ROS) system [53]. Miele et al showed in L6 skeletal muscle cells that AGEs affect glucose metabolism by impairing insulin-induced insulin receptor substrate (IRS) signaling through protein kinase Cα-mediated mechanism [37]. The same research group also showed in the muscle cells that methylglyoxal, an essential source of intracellular AGEs, hampers a key insulin signaling molecule [54]. Recent observations by Unoki et al also showed that AGEs impair insulin signaling in adipocytes by increasing generation of intracellular ROS [55]. Thus, AGEs may not only induce the debilitating complications of diabetes, but may also contribute to the impairment of insulin signaling in insulin-target tissues which could be involved in pathophysiology of insulin resistance, metabolic syndrome and diabetes.

3.4 Endogenous RAGE ligands, insulin resistance and metabolic syndrome

RAGE also interacts with other endogenous non-glycated peptide ligands including S100/calgranulin [56], amphoterin (also termed as high mobility group box 1 protein, HMGB1) [57, 58], amyloid fibrills [59], transthyretin [60], and a leukocyte integrin, Mac-1 [61], many of which are important inflammatory regulators. Some of these inflammatory ligands for RAGE may be involved in pathogenesis of obesity and metabolic syndrome. Early studies show expression of S100B protein in pre- and mature- adipocyte and is induced during adipogenesis [62, 63]. Physiological S100B levels appear to closely reflect adipose tissue mass or insulin resistance in humans [64-66]. HMGB1 is also found to be expressed in human adipose tissue with the expression level associated with the fat mass and obesity-associated gene [67]. Moreover, growing evidences suggest that infiltration of inflammatory cells, including macrophages, play fundamental roles in adiposity and metabolic syndrome [68-70]. MAC-1, an integrin expressed in macrophage, can act as a RAGE ligand [61], and may be involved in adipogenesis through interaction with RAGE.

4. RAGE and its potential link with insulin resistance and metabolic syndrome

4.1 Structure and function of RAGE

RAGE is a multiligand cell-surface protein that was isolated from bovine lung in 1992 by the group of Schmidt and Stern [71, 72]. RAGE belongs to the immunoglobulin superfamily of cell surface molecules and has an extracellular region containing one "V"-type immunoglobulin domain and two "C"-type immunoglobulin domains [71, 72] (Figure 1). The extracellular portion of the receptor is followed by a hydrophobic trans-membrane-spanning and then by a highly charged, short cytoplasmic domain which is essential for intracellular RAGE signaling. RAGE is initially identified as a receptor for CML-modified proteins [73], a major AGE in vivo [74]. Three-dimensional structure of the recombinant AGE-binding domain by using multidimensional heteronuclear NMR spectroscopy revealed that the domain assumes a structure similar to those of other immunoglobulin V-type domains [75, 76]. Three distinct surfaces of the V domain were identified to mediate AGE-V domain interactions [75]. The site-directed mutagenesis studies identified the basic amino acids which play a key role in the AGE binding activities [76]. As mentioned in the previous sentence, RAGE also interacts with other endogenous non-glycated peptide ligands, many of which are important inflammatory regulators. The common characteristics of these ligands are the presence of multiple β-sheets [61, 77, 78]. RAGE is thought to interact with these ligands through their shared three-dimensional structure.

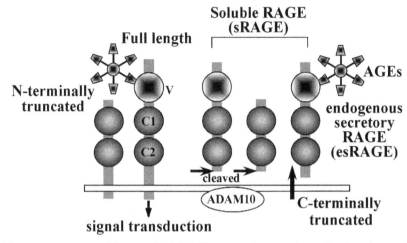

Fig. 1. Numerous truncated forms of RAGE. There are three major spliced variants of RAGE: full length, N-terminally truncated, and C-terminally truncated. The C-terminally truncated form of RAGE is secreted from the cell and is named endogenously secreted RAGE (esRAGE). esRAGE has a V-domain, which is essential for binding with ligands, and is capable of competing with RAGE signaling as a decoy receptor. There are other forms of soluble RAGE (sRAGE) that are cleaved from cell-surface RAGE by matrix metalloproteinases. The ELISA assay for sRAGE measures all soluble forms including esRAGE in human plasma, while the ELISA for esRAGE measures only esRAGE, using polyclonal antibody raised against the unique C-terminus of the esRAGE sequence.

4.2 Inflammatory signaling mediated by RAGE

Ligand engagement of RAGE leads to prolonged inflammation, resulting in a RAGE-dependent expression of proinflammatory mediators such as monocyte chemoattractant protein-1 (MCP-1) and vascular cell adhesion molecule-1 (VCAM-1) [79, 80]. RAGE-mediated proinflammatory signals could potentially converge with insulin signaling system (Figure 2). The engagement of RAGE has been reported to induce activation of the transcription factor nuclear factor-κB (NF-κB). Recent reports by Harja et al demonstrate that RAGE mediates upregulation of VCAM-1 in response to S100b and oxLDL and JNK MAP kinase underlies the RAGE ligand-stimulated molecular events [81]. It is not known at present whether this is also the case in classical insulin target cells. JNK activity is strikingly increased in critical metabolic sites (eg. adipose and liver tissues) [82], and is shown to be crucial in IRS-1 phosphorylation and consequently insulin resistance [82, 83]. Moreover, the main pathological consequence of RAGE ligation is the induction of intracellular reactive oxygen species (ROS) via NAD(P)H oxidases and other identified mechanisms such as mitochondrial electron transport chain [84], which consequently results in oxidative stress in the cells [85]. Oxidative stress is emerging as a feature of obesity and an important factor in the development of insulin resistance [86, 87]. Both the NF-kB and JNK pathways can be activated under the conditions of oxidative stress, and this may be important for the ability of ROS to mediate insulin resistance. RAGE has a short cytosolic portion that contains 43 amino acids [72]. So far, adaptors and/or scaffold proteins that interact with the cytosolic tail of RAGE has barely been identified. The RAGE mutant lacking the 43-residue C-terminal tail fails to activate NF-κB, and expression of the mutant receptor results in a dominant negative effect against RAGE-mediated production of proinflammatory cytokines from macrophages [56, 57].

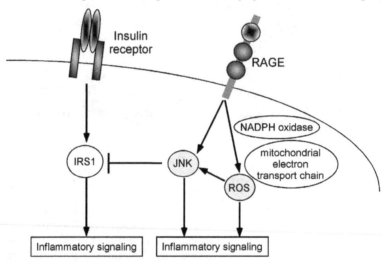

Fig. 2. RAGE and insulin signaling. RAGE is known to activate JNK pathway, which could phosphorylate serine-residue of insulin receptor substrate (IRS) and inhibit its activity. RAGE mediated generation of reactive oxygen spices (ROS) may alternatively influence insulin signaing.

4.3 RAGE and obesity

A.   RAGE, adiposity and atherosclerosis in mouse model

Recent reports suggest that RAGE could be involved in progression of obesity. Recent study in humans shows RAGE mRNA expression in subcutaneous adipose tissues [88]. Although this study does not delineate which cells in adipose tissue express RAGE, our current animal study shows RAGE expression in adipocyte as wells as endothelial cells in adipose tissues [16]. We have shown by using apo E/RAGE double knockout mice that progression of atherosclerosis is closely associated with RAGE-regulated adiposity in non-diabetic conditions [16]. As shown in Figure 3, apoE-/-RAGE-/- mice fed either with standard or atherogenic diet exhibited significantly decreased atherosclerotic plaque area in aorta as compared with apoE-/-RAGE+/+ mice. Importantly, apoE-/-RAGE-/- mice also exhibited significantly less body weight, epididymal fat weight and epididymal adipocyte size than apoE-/-RAGE+/+ mice at 20-weeks of age (Figure 4). Decreased body weight, epididymal fat weight, and adipocyte size are associated with higher plasma adiponectin levels and decreased atherosclerosis progression. RAGE is involved in adiposity even in apo E+/+ genetic background. At 20-weeks of age, epididymal adipocyte size of RAGE-/- mice was significantly smaller than that of RAGE+/+ mice (data not shown).

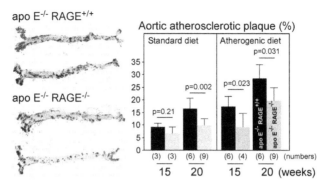

Fig. 3. RAGE deficiency suppresses atherosclerotic progression in apoE deficient mice. Representative aortas from apoE-/-RAGE+/+ and apoE-/-RAGE-/- mice (20-weeks old) fed with atherogenic diet were shown in left panel. Right panel summarizes the quantitative analyses. Plaque area was represented as percentages of the total plaque area. Columns represent mean ± standard deviation. Black columns represent apoE-/-RAGE+/+ mice, and grey columns, apoE-/-RAGE-/- mice. P values were analyzed by Student's unpaired t-test. Reproduced from ref [16].

B. Roles of inflammatory cells?

RAGE is also known to play fundamental role in functions of inflammatory cells [61, 89, 90], raising an intriguing possibility that RAGE's function on adiposity may be mediated through its function in inflammatory cells infiltrated in adipose tissues. In our study in apoE-/- genetic background fed with atherogenic diet, numbers of Mac-3-positive inflammatory cells infiltrated in the epididymal adipose tissues of RAGE+/+apoE-/- mice and RAGE-/-apoE-/- did not show significant differences, and crown-like structure were barely detected in epididymal adipose tissue in both groups even at 20-week of age. In standard diet-fed mice, even though the adiposity was significantly different between RAGE+/+apoE-/- and RAGE-/-apoE-/- mice, crown-like structure were not detected in epididymal adipose tissues in both groups even at 20-week of age. Further in apoE+/+ genetic background at 10 week of age when significantly different pattern of gene expression was observed between WT and RAGE-/- mice, no marked differences in expressions of macrophage markers were observed as analyzed by gene microarray. At that age, macrophage infiltration in adipose tissues is also reported to be scant [91]. Thus, it appears infeasible to RAGE acting primarily at inflammatory cells at least in early phase of adiposity, while RAGE expressed in endothelial cells or adipocyte might play fundamental roles.

C. RAGE-regulated genes in adipose tissue: gene chip analysis

To explore potential mechanisms underlying RAGE-regulation of adiposity, mRNA expression profile in epididymal adipose tissue was compared between RAGE+/+ and RAGE-/- mice using Affymetrix GeneChip Mouse Genome 430 2.0. We isolated total RNA from epididymal adipose tissue at 10-weeks of age, at which phenotypic change in adipocyte size was not observed. Using 3 μg of total RNA, 59.8% and 61.4% of 45,037 genes were revealed to be present in RAGE+/+ and RAGE-/- adipose tissue, respectively. Comparison analysis of the genes (RAGE+/+ adipose tissue as base line) revealed that 10.3% of the total genes were decreased, while 11.7% increased in RAGE-/- adipose tissue. As compared with RAGE+/+ adipose tissue, 623 genes were downregulated to less than a half, and 2,470 genes upregulated more than 2 fold in RAGE-/- adipose tissue.

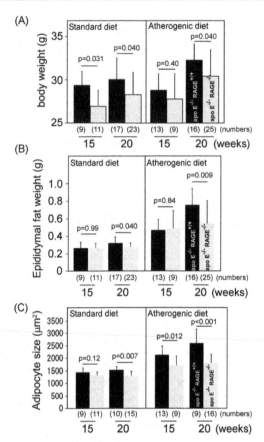

Fig. 4. RAGE deficiency is associated with decreased body weight, epididymal fat weight and adipocyte size in apolipoprotein E (apoE)-deficient genetic background. (A) Comparisons of body weight between apoE-/-RAGE+/+ and apoE-/-RAGE-/- mice fed with standard or atherogenic diet. (B) Comparisons of epididymal fat weight between apoE-/-RAGE+/+ and apoE-/-RAGE-/- mice fed with standard or atherogenic diet. (C) Comparisons of adipocyte size in epididymal adipose tissues. Columns represent mean ± standard deviation. P values were analyzed by Student's t-test. Modified from ref [16].

D.   RAGE-regulated genes in adipose tissue: ontology analysis

To mine specific group of genes involved in adiposity regulated by RAGE, gene ontology analyses were performed. Downregulated genes in RAGE-/- adipose tissue were significantly accumulated in the ontology terms of metabolic process including acetyl-CoA biosynthetic process, neutral lipid biosynthetic process, pyruvate metabolic process, gluconeogenesis, glycogen biosynthetic process, and NADPH regeneration. Interestingly, genes involved in fat cell differentiation were also identified to be accumulated as down-regulated in RAGE-/- adipose tissue. Ontology terms of glucose transport and neutral amino acid transport were also significantly extracted as downregulated in RAGE-/- adipose tissue. Insulin receptor signaling pathway was a highly significant ontology term downregulated in RAGE-/- adipose tissue. On the contrary, many of the genes upregulated in RAGE-/- adipose tissue were accumulated in ontology terms including cell adhesion, endocytosis, T cell activation, prostaglandin biosynthesis, protein binding, protein folding, processing and glycoprotein biosynthetic process, many of which are known be associated with cellular mechanisms for inflammation and defensive process. Nitrogen compound metabolic process, including amino acid metabolic process, was also identified to be a significant ontology term upregulated in RAGE-/-. Interestingly, upregulated genes in RAGE-/- tissue were also significantly accumulated in ontology term for cell redox homeostasis process.

E.   RAGE-regulated genes in adipose tissue: pathway analysis

To further identify potential pathways involved in RAGE-regulation of adiposity, KEGG pathway analyses were performed (Table 1). In accordance with the ontology analyses, insulin signaling pathway, pyruvate metabolism, fatty acid biosynthesis and gluconeogenesis were identified to be downregulated pathways in RAGE-/- adipose tissue. PPAR signaling and adipocytokine signaling were also identified to be downregulated in RAGE-/- adipose tissue. Similar to gene ontology analyses, inflammatory pathways including cell adhesion molecules and leukocyte transendothelial migration were the significant pathways upregulated in RAGE-/- mice. Pathways including amino acid metabolic pathways, nitrogen metabolism, glycan biosynthesis, structure and degradation were the pathways significantly upregulated in RAGE-/- adipose tissues.

WT>RAGE-/- (>= 3 fold)	count	P value
Insulin signaling pathway	4/137	0.0002
Fatty acid biosynthesis	1/6	0.0146
ErbB signaling pathway	2/87	0.0179
Ethylbenzene degradation	1/10	0.0242
1- and 2-methylnaphthalene degradation	1/20	0.0478
Jak-STAT signaling pathway	2/151	0.0500
WT>RAGE (>= 2 fold)		
Alanine and aspartate metabolism	7/33	<0.0001
Pyruvate metabolism	7/41	<0.0001
Insulin signaling pathway	12/137	0.0001
Adipocytokine signaling pathway	6/71	0.0059
Fatty acid biosynthesis	2/6	0.0077
Glycerophospholipid metabolism	5/61	0.0134
Glycerolipid metabolism	4/41	0.0148
Glutathione metabolism	4/42	0.0160
Ethylbenzene degradation	2/10	0.0216
Valine, leucine and isoleucine biosynthesis	2/10	0.0216
Type II diabetes mellitus	4/46	0.0218
Metabolism of xenobiotics by cytochrome P450	5/71	0.0245
Sulfur metabolism	2/11	0.0260
PPAR signaling pathway	5/74	0.0287
Propanoate metabolism	3/30	0.0320
Glycolysis / Gluconeogenesis	4/53	0.0345
Circadian rhythm	2/13	0.0358

Table 1. Pathway analyses of the genes differentially expressed in WT vs. RAGE-/- epididymal adipose tissue.

WT<RAGE-/- (>= 2 fold)		
Arginine and proline metabolism	6/33	0.0008
Glycine, serine and threonine metabolism	7/47	0.0010
Glycerolipid metabolism	5/41	0.0128
N-Glycan degradation	3/15	0.0136
Cyanoamino acid metabolism	2/6	0.0163
One carbon pool by folate	3/16	0.0164
Glycosphingolipid biosynthesis - ganglioseries	3/16	0.0164
Polyunsaturated fatty acid biosynthesis	3/17	0.0194
Ether lipid metabolism	4/32	0.0234
Cell adhesion molecules (CAMs)	10/147	0.0301
Prostate cancer	7/88	0.0317
Nitrogen metabolism	3/21	0.0343
Glycosylphosphatidylinositol(GPI)-anchor biosynthesis	3/21	0.0343
Glycan structures - biosynthesis 1	8/114	0.0428
WT<RAGE-/- (>= 3 fold)		
Arginine and proline metabolism	4/33	0.0053
Polyunsaturated fatty acid biosynthesis	3/17	0.0054
Glycerolipid metabolism	4/41	0.0114
Leukocyte transendothelial migration	7/115	0.0118
Glutathione metabolism	4/42	0.0124
Cell adhesion molecules (CAMs)	8/147	0.0137
O-Glycan biosynthesis	3/27	0.0199
Thyroid cancer	3/28	0.0219
Glyoxylate and dicarboxylate metabolism	2/14	0.0358
Glycan structures - biosynthesis 1	6/114	0.0364
One carbon pool by folate	2/16	0.0459
Pantothenate and CoA biosynthesis	2/16	0.0459

Table 1. Pathway analyses of the genes differentially expressed in WT vs. RAGE-/-
epididymal adipose tissue (continuation).

F.   RAGE-regulated genes in adipose tissue: real time RT-PCR confirmation

Adipogenesis related genes including, lipin 1, peroxisome proliferator-activated receptor
(PPAR)-γ, adipose differentiation related protein, were shown to be downregulated in
RAGE-/- mice. Fatty acid binding protein 5, 1-acylglycerol-3-phosphate O-acyltransferase 2,
diacylglycerol O-acyltransferase 2, monoacylglycerol O-acyltransferase 1, acetoacetyl-CoA
synthetase, acetyl-coenzyme A carboxylase α were downregulated in RAGE-/- adipose
tissue, which could be an essential mechanisms for decreased adiposity in RAGE-/- mice. In
insulin signaling, phosphatidylinositol 3-kinase (p85α), adaptor protein with pleckstrin
homology and src (APS), sorbin and SH3 domain containing 1 (CAP), insulin receptor
substrate (IRS) 1 and 3, thymoma viral proto-oncogene 2 / similar to serine/threonine
kinase (Akt), Protein phosphatase 1 regulatory (inhibitor) subunit 3C, facilitated glucose

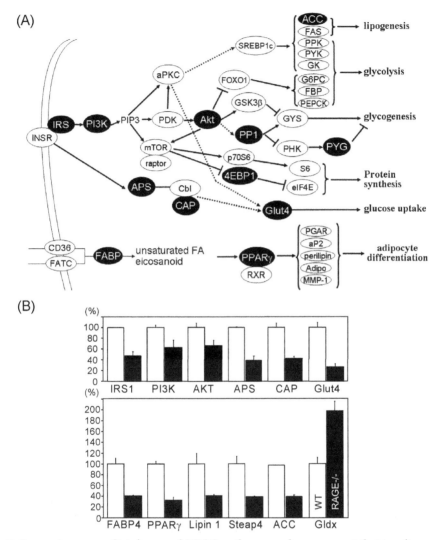

Fig. 5. Gene microarray, Ontology and KEGG pathway analyses suggest that insulin signaling and adipocyte differentiation are the potential pathways regulated by RAGE. (A) Figure summarizes the results of Ontology and KEGG pathway analyses. Genes suppressed in RAGE-/- adipose tissue were described in black circles. (B) Changes in mRNA expression obtained by gene microarray analyses were confirmed by real-time quantitative RT-PCR analyses. All changes in gene expression were statistically significance (p<0.05, Student's t-test). IRS-1: insulin receptor substrate 1, PI3K: phosphatidylinositol 3-kinase (p85α), AKT: thymoma viral proto-oncogene 2 / similar to serine/threonine kinase, APS: adaptor protein with pleckstrin homology and src, CAP: sorbin and SH3 domain containing 1, Glut4: facilitated glucose transporter member 4, FABP4: fatty acid binding protein 4, PPAR-γ: peroxisome proliferator-activated receptor, Steap4: six-transmembrane epithelial antigen of prostate 4, ACC: acetyl-coenzyme A carboxylase α, Gldx: glutaredoxin.

transporter member 4 (Glut 4) were identified to be downregulated in RAGE-/- adipose tissue. Figure 5A shows genes specifically suppressed in RAGE-/- adipose tissue (closed circles) in insulin signaling and adipocyte differentiation pathways. Real-time quantitative RT-PCR analyses confirmed the genes in the pathways were indeed down-regulated in RAGE-/- adipose tissue (Figure 5B). These results altogether suggest direct role of RAGE in adiposity. Although in which cell types RAGE is principally working, insulin signaling and adipocyte signaling pathway in adipose tissue appear to play important part in RAGE regulation of adiposity.

4.4 RAGE, endothelial dysfunction and insulin resistance

Impaired insulin action, when assessed by fasting serum insulin levels or the homeostasis model assessment of insulin resistance (HOMA-IR) [92], is associated with atherosclerosis and an increased risk of myocardial infarction. Insulin resistance is associated with endothelial dysfunction [93] and may serve as a link between insulin resistance and atherosclerosis. Recent findings by Harja et al highlighted the involvement of RAGE in endothelial dysfunction [81]. Endothelium-dependent vasorelaxation was tested in isolated mouse aortic rings from *apoE-/-* and *apoE-/-RAGE-/-* mice, and relaxation response to acetylcholine was significantly improved in the RAGE deficient mouse. Similarly, impaired endothelial function in diabetic obese mice was also shown to be mediated by AGEs/RAGE system, since blockade of AGE-RAGE interaction by soluble RAGE significantly improved endothelial function [94]. Recent clinical observations by Linden et al [44] also implies AGEs/RAGE system is involved in impaired endothelial function in patients with chronic kidney diseases. Thus, not only by the interaction at the cellular signaling level, but RAGE appears to impair endothelial function and potentially blood flow in insulin target tissues, leading to insulin resistance in vivo.

5. C-terminally truncated form of RAGE (soluble RAGE, sRAGE) as potential biomarkers for cardiovascular diseases, metabolic syndrome and insulin resistance

5.1 Truncated form of RAGE

Numerous truncated forms of RAGE have recently been described [12, 95-98] (Figure 1). Two major spliced variants of RAGE mRNA, N-terminal and C-terminal truncated forms, have been most extensively characterized [12]. The N-truncated isoform of RAGE mRNA codes for a 303-amino-acid protein lacking the N-terminal signal sequence and the first V-like extracellular domain. The N-truncated form is incapable of binding with AGEs, since the V-domain is critical for binding of the ligand [71]. The N-truncated form of RAGE appears to be expressed on the cell surface similar to the full-length RAGE, although its biological roles remain to be elucidated [99]. It has been suggested that this form of RAGE could be involved in angiogenic regulation in a fashion independent of the classical RAGE signaling pathway [99].

5.2 Endogenous secretory RAGE (esRAGE)

The C-terminal truncated form of RAGE lacks the exon 10 sequences encoding the transmembrane and intracytoplasmic domains [12]. This spliced variant mRNA of RAGE encodes a protein consisting of 347 amino acids with a 22-amino-acid signal sequence, and is released from cells. This C-truncated form is now known to be present in human circulation and is named endogenous secretory RAGE (esRAGE) [12]. Regulation of alternative splicing of the RAGE is recently shown to be regulated through G-rich cis-elements and heterogenous nuclear ribonucleoprotein H [100]. esRAGE was found to be capable of neutralizing the effects of AGEs on endothelial cells in culture [12]. Adenoviral overexpression of esRAGE in vivo in mice reverses diabetic impairment of vascular dysfunction [101]. Thus, the decoy function of esRAGE may exhibit a feedback mechanism by which esRAGE prevents the activation of RAGE signaling.

5.3 Soluble RAGE generated by shedding

It has also been suggested that some sRAGE isoforms that could act as decoy receptors may be cleaved proteolytically from the native RAGE expressed on the cell surface [102], suggesting heterogeneity of the origin and nature of sRAGE. This proteolytic generation of sRAGE was initially described as occurring in mice [103]. Recent studies suggest that ADAM10 and MMP9 to be involved in RAGE shedding [13, 14]. ADAM is known as a shedase to shed several inflammatory receptors and can be involved in regulation of RAGE/sRAGE balance. A RAGE gene polymorphism is shown to be strongly associated with higher sRAGE levels, although the mechanism by which the polymorphism alters the sRAGE levels remains to be elucidated [104]. Thus, the molecular heterogeneity of the diverse types of sRAGE in human plasma could exert significant protective effects against RAGE-mediated toxicity. However, the endogenous action of sRAGE may not be confined to a decoy function against RAGE-signaling. In HMGB1-induced arthritis model, for example, sRAGE is found to interact with Mac-1, and act as an important proinflammatory and chemotactic molecule [105]. Further analyses are warranted to understand more about the endogenous activity of sRAGE.

5.4 Circulating sRAGE and esRAGE in diseases

A.   Circulating sRAGE and cardiovascular diseases

Since sRAGE and esRAGE may be involved in feedback regulation of the toxic effects of RAGE-mediated signaling, recent clinical studies have focused on the potential significance of circulating sRAGE and esRAGE in a variety of pathophysiological conditions, including atherosclerotic disorders, diabetes, hypertension, Alzheimer's dieases and chronic kidney diseases (Table 2). First, Falcone et al [106] reported that total sRAGE levels are significantly lower in patients with angiographically proven coronary artery disease (CAD) than in age-matched healthy controls. The association between circulating sRAGE and angiographic observations was shown to be dose-dependent, with individuals in the lowest quartile of sRAGE exhibiting the highest risk for CAD. Importantly, this cohort consisted of a non-diabetic population, suggesting that the potential significance of sRAGE is not confined to diabetes. Falcone et al also showed that the association between sRAGE and the risk of CAD was independent of other classical risk factors. Their findings are reproduced later by several research groups in larger numbers of subjects, and are also extended to other atherosclerotic diseases, such as carotid atherosclerosis, cerebral ischemia, and aortic valve stenosis (Table 2). Patients with

Alzheimer disease have also lower levels of sRAGE in plasma than patients with vascular dementia and controls, suggesting a role for the RAGE axis in this clinical entity as well [107].

sRAGE		references
CAD (non-DM)	decreased	106, 153, 154
	increased	124
Calcified aortic valve stenosis	decreased	155
Carotid atherosclerosis	decreased	156
Cerebral ischemia	decreased	157 158 127
Alzheimer's disease	decreased	107 159
Endothelial dysfunction	decreased	160
Diabetes (type 1)	increased	122
Diabetes (type 2)	increased	123, 124
	decreased	120, 121
Hypertension	decreased	117
NASH	decreased	118 119
Chronic kidney disease	increased	109, 123, 129, 161
		162
Oxidative stress and inflammatory markers	positively associated	163, 164
	inversely associated	121
esRAGE		
Insulin resistance	inversely associated	15
Metabolic syndrome	decreased	15
Diabetes (type 1)	decreased	108, 110
Diabetes (type 2)	decreased	15, 113
Hypertension	decreased	15
NASH	decreased	119
Carotid atherosclerosis	decreased	15, 110-112
	no association	109
CAD	decreased	113, 153
Altzheimer's disease	decreased	165
Chronic kidney disease	increased	114, 161 162

Table 2. Levels of circulating soluble RAGE in cardiovascular and metabolic diseases.

B. Circulating esRAGE and cardiovascular diseases

Following development of an ELISA system to specifically measure human esRAGE [108], we measured plasma esRAGE level and cross-sectionally examined its association with atherosclerosis in 203 type 2 diabetic and 134 non-diabetic age- and gender-matched subjects [15]. esRAGE levels were inversely correlated with carotid and femoral atherosclerosis, as measured as intimal-medial thickness (IMT) by arterial ultrasound. Stepwise regression analyses revealed that plasma esRAGE was the third strongest and an independent factor associated with carotid IMT, following age and systolic blood pressure[15]. Importantly however, when non-diabetic and diabetic groups were separately analyzed, inverse correlation between plasma esRAGE level and IMT was significant in non-diabetic population only, suggesting a potential significance of esRAGE in non-diabetic condition. No association of plasma esRAGE with IMT in diabetes was also reported in other study with 110 Caucasian type 2 diabetic subjects [109]. Another Japanese research group found an inverse correlation between plasma esRAGE and carotid atherosclerosis in type 1 [110] and type 2 diabetic subjects [111]. Recently, the same research group also longitudinally examined the predictive significance of plasma esRAGE and sRAGE on progression of carotid atherosclerosis, and found that low circulating esRAGE level as well as sRAGE level was an independent risk factor for the progression of carotid IMT in type 1 diabetic subjects [112]. In Chinese type 2 diabetic patients, plasma esRAGE is recently shown to be decreased in angiographically-proved patients with coronary artery disease than those without it [113].

C. Low circulating sRAGE as a predictor of cardiovascular diseases

We also reported an observational cohort study in patients with end-stage renal disease (ESRD) and longitudinally evaluated the effect of plasma esRAGE on cardiovascular mortality [114]. The cohort in that study included 206 ESRD subjects, who had been treated by regular hemodialysis for more than 3 months. Even though the plasma esRAGE levels at baseline were higher in ESRD subjects than in those without kidney disease, the subjects in the lowest tertile of plasma esRAGE levels exhibited significantly higher cardiovascular mortality, but not non-cardiovascular mortality. Importantly, even in the subpopulation of non-diabetic subjects alone, low circulating esRAGE level was a predictor of cardiovascular mortality, independent of the other classical risk factors. Thus, low circulating esRAGE or sRAGE level is a potential predictor for atherosclerosis and cardiovascular diseases even in non-diabetic population.

D. Circulating sRAGE, esRAGE and metabolic syndrome

Several components of metabolic syndrome have been shown to be associated with altered plasma sRAGE or esRAGE levels. We first reported that plasma esRAGE levels are already decreased in patients with impaired glucose tolerance as compared with those with normal glucose tolerance (Figure 6A). Moreover, patients with metabolic syndrome showed significantly lower plasma esRAGE than those without it (Figure 6A). Plasma esRAGE levels are inversely correlated with many of the components of metabolic syndrome including body mass index (Figure 6B), blood pressures, fasting plasma glucose, serum triglyceride, and lower HDL-cholesterol levels [15]. The majorities of these correlations remained significant even when the non-diabetic or type 2 diabetic subpopulation was extracted for analyses. An inverse correaltion between esRAGE (or sRAGE) and body mass index was also found for control subjects [115], those with type 1 diabetes [116], and those with ESRD [114]. Patients with hypertension have been found to have lower plasma sRAGE or esRAGE levels [15, 117]. Importantly, our findings also showed that plasma esRAGE was also inversely associated with insulin resistance index, HOMA (Figure 6B), suggesting esRAGE and sRAGE as potential biomarkers for metabolic syndrome and insulin resistance, which could be associated with altered cardiovascular outcomes. Both sRAGE and esRAGE are found to be decreased in patients with liver steatosis [118, 119], which is know to be deeply associated with visceral fat accumulation and insulin resistance.

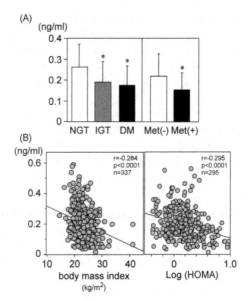

Fig. 6. Plasma esRAGE levels are decreased in glucose intolerance, metabolic syndrome, obesity and insulin resistance (A) Left panel demonstrates the levels of plasma esRAGE in subjects with normal glucose tolerance (NGT) (n=118), impaired glucose tolerance (IGT) (n=16), and type 2 diabetes (DM) (n=203). Right panel compares the plasma esRAGE levels in subjects with (n-53) or without (n=282) metabolic syndrome (Met) as characterized by modified NCEP criteria. * p<0.05, ANOVA with multiple comparison (Scheffe's type). (B) Plasma esRAGE levels were inversely associated with body mass index or HOMA insulin resistance index. Logarithm-transformed HOMA index was used for the analyses because of the skewed distribution. Modified from ref [15].

E. Circulating sRAGE and esRAGE in diabetes

The findings regarding plasma levels of the soluble form of RAGE in diabetes are quite confusing. We and other groups have found that plasma esRAGE level is significantly lower in type 1 and type 2 diabetic patients than in non-diabetic controls [15, 110]. Plasma sRAGE levels have also been shown to be decreased in diabetic subjects [120, 121], although conflicting findings have also been reported for type 1 [122] and type 2 diabetes [123, 124]. We examined plasma sRAGE levels by different ELISA system using esRAGE as a standard protein and different sets of antibodies against whole RAGE molecule [125]. In our hand, type 2 diabetic subjects without overt nephropathy (0.60 ± 0.28 ng/ml) exhibited significantly (p<0.001, Student's t-test) lower plasma sRAGE level than non-diabetic controls (0.77 ± 0.34 ng/ml) [11]. Of note, when diabetic subjects alone were extracted for analyses, a direct association was not observed between plasma soluble RAGE (both sRAGE and esRAGE) levels and the status of glycemic control (i.e. glycated hemoglobin A1c) [15, 109, 116, 120, 126]. Thus, these complex findings in diabetic subjects suggest that levels of plasma soluble forms of RAGE are not determined simply by status of glycemic control, and that even plasma esRAGE and sRAGE levels may be under the control of distinct mechanisms. Recent study suggests that sRAGE levels may be significantly influenced by ethnicity [127], which may partially explain controversial findings.

F.   Circulating sRAGE and esRAGE in CKD

Another important component that can affect plasma sRAGE is the presence of chronic kidney disease. It has been shown that, in peripheral monocytes from subjects with varying severities of CKD, RAGE expression is closely associated with worsening of CKD and is strongly correlated with plasma levels of pentosidine, a marker for AGEs [128]. Circulating sRAGE levels have been shown to be increased in patients with decreased renal function, particularly those with ESRD [109, 123, 129]. Our observations revealed that plasma esRAGE levels in type 2 diabetic subjects without CKD are lower than non-diabetic controls, which is gradually elevated in accordance with progression of CKD [11]. Plasma sRAGE levels in diabetic subjects without CKD also exhibited significantly lower than those of non-diabetic controls [11]. Thus, plasma sRAGE and esRAGE are markedly affected by the presence of CKD, which might make the interpretation of the role of soluble RAGE quite complicated [130]. It remains to be determined whether the increase in plasma esRAGE in CKD is caused by decreased renal function alone or whether esRAGE levels are upregulated to protect against toxic effects of the RAGE ligands. Successful kidney transplantation resulted in significant decrease in plasma sRAGE [131], implying that the kidneys play a role in sRAGE removal.

6. RAGE and Soluble RAGE as a therapeutic target against metabolic syndrome, insulin resistance and cardiovascular disease?

6.1 Soluble RAGE as a therapeutic tool in animal disease models

Potential usefulness of soluble RAGE for prevention and treatment of inflammatory diseases has been demonstrated in many animal models. Blockade of RAGE by administration of genetically engineered sRAGE successfully prevented the development of micro- [132, 133] and macrovascular complications in diabetes [134-136]. We have also shown that adenoviral overexpression of esRAGE successfully restored the impaired angiogenic response in diabetic mice [101]. Sakaguchi et al found that administration of sRAGE markedly suppressed neointimal formation following arterial injury in non-diabetic mice [137]. Soluble RAGE has also been shown to effectively prevent the development of diabetes [138], protect against tumor growth and metastasis [58], improve the outcome of colitis [56], restore impaired wound healing [139], and suppress Alzheimer disease-like conditions [140]. These effects of soluble RAGE in animal models could be explained by its decoy function, inhibiting RAGE interaction with its proinflammatory ligands, which might be applicable to human diseases as well. Since our findings strongly suggest the role of RAGE in adiposity, metabolic syndrome and atheroslcerosis [16], RAGE/soluble RAGE axis could also be a potential therapeutic target against these pathophysiological conditions.

6.2 Potential regulatory mechanisms of circulating soluble RAGE

So far, limited findings are available regarding the mechanisms of regulation of circulating esRAGE or sRAGE in humans. A tissue microarray technique using a wide variety of adult normal human preparations obtained from surgical and autopsy specimens revealed that esRAGE was widely distributed in tissues, including vascular endothelium, monocyte/macrophage, pneumocytes, and several endocrine organs [141]. However, it is unclear at present from which organ or tissue plasma sRAGE or esRAGE originate. Circulating AGEs may be involved in regulation of the secretion or production of soluble RAGE, since AGEs are known to upregulate RAGE expression in vitro [142]. esRAGE could be simultaneously upregulated by AGEs and act as a negative feedback loop to compensate for the damaging effects of AGEs. We and others have found positive correlations between plasma sRAGE or esRAGE and AGEs [11, 114-116, 123]. Significant positive correlation between plasma esRAGE and pentosidine was observed both in hemodialysis and non-hemodialysis subjects [11]. However, plasma CML did not significantly correlated with plasma esRAGE both in hemodialysis and non-hemodialysis subjects. AGEs-mediated regulation of soluble RAGE is also supported by the findings that the suppression of sRAGE expression in diabetic rat kidney is reversed by blockade of AGEs accumulation with alagebrium [143]. Other inflammatory mediators, such as S100, tumor necrosis factor α, and C-reactive protein, could also be potential candidates for regulation of the plasma level of soluble RAGE in humans [120, 142, 144]. Moreover, Geroldi et al [145] showed that high serum sRAGE is associated with extreme longevity, suggesting that understanding the intrinsic regulation of RAGE and soluble RAGE is important for longevity/anti-aging strategies. Without doubt, further understanding of the regulation of soluble RAGE will be most helpful in delineating potential targets for therapeutic application of soluble RAGE.

6.3 Pharmacological agents regulate circulating sRAGE and esRAGE

A. Angiotensin-converting enzyme inhibitor

It would be essential to determine whether currently available pharmacological agents can regulate plasma sRAGE or esRAGE. Potential agents that may affect circulating soluble RAGE include the angiotensin-converting enzyme (ACE) inhibitor [146], thiazolidinediones (TZD) [147] and statins [148-150], which are known to modulate the AGEs-RAGE system in culture. Forbes et al [146] showed that inhibition of angiotensin-converting enzyme (ACE) in rats increased renal expression of sRAGE, and that this was associated with decreases in expression of renal full-length RAGE protein. They also showed that plasma sRAGE levels were significantly increased by inhibition of ACE in both diabetic rats and in human subjects with type 1 diabetes. Thus, one attractive scenario is that the protective effect of ACE inhibition against progression of renal dysfunction is mediated through regulation of RAGE versus soluble RAGE production.

B. Statin

Tam et al recently reported changes in serum levels of sRAGE and esRAGE in archived serum samples from a previous randomized double-blind placebo-controlled clinical trial that explored the cardiovascular effects of atorvastatin in hypercholesterolemic Chinese type 2 diabetic patients, and found that atorvastatin can increase circulating esRAGE levels [150].

C. Thiazolidinedione

For thiazolidinedione, a randomised, open-label, parallel group study was performed with 64 participants randomised to receive add-on therapy with either rosiglitazone or sulfonylurea to examine the effect on plasma soluble RAGE [151]. At 6 months, both rosiglitazone and sulfonylurea resulted in a significant reduction in HbA1c, fasting glucose and AGE. However, significant increases in total sRAGE and esRAGE were only seen in the rosiglitazone group. In a recent study in type 2 diabetes mellitus patients, pioglitazone, but not rosiglitazone, significantly raised sRAGE levels [152], suggesting that all thiazolidinedione may not act similarly. Nevertheless, thiazolidinedione could be one promising candidate which increase circulating levels of esRAGE and sRAGE, and RAGE/soluble RAGE regulation may be involved in thiazolidinedione-mediated improvement of insulin resistance. Finally, we have started the randomized clinical trial comparing the effect of pioglitazone with glimepiride on plasma sRAGE and esRAGE, expression of RAGE on peripheral mononuclear cells, and RAGE shedase gene expression in type 2 diabetic patients (UMIN000002055). This study will be of particular importance to understand the regulatory mechanisms of sRAGE and esRAGE in clinical setting.

7. Summary

The findings discussed here implicated pivotal role of RAGE system in initiation and progression of metabolic syndrome, insulin resistance and atherosclerosis. Provided that continuous RAGE activation represents the concept of "metabolic memory", metabolic syndrome might be conceptualized as memorized long-term subtle inflammation and oxidative stress using RAGE as an inflammatory scaffold. In this system, endogenous inflammatory RAGE ligands may be profoundly involved (Figure 7). Further, sRAGE or esRAGE could serve as a biomarker as well as a therapeutic target for these disease conditions. Obviously there are many missing parts to be veiled to further understand the role of RAGE/soluble RAGE axis in metabolic syndrome and insulin resistance. However, we believe our findings and this concept would open up a new research field which could further precede our understanding of the RAGE biology.

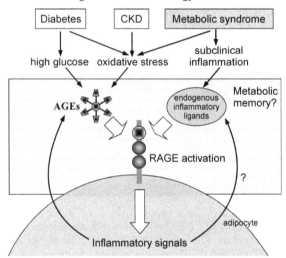

Fig. 7. Metabolic syndrome may be an aspect of "metabolic memory" conceptualized as prolonged RAGE activation through subclinical information.

8. Acknowledgments

The authors thank all colleagues in the Hyogo College of Medicine, Osaka City University Graduate School of Medicine, and Kanazawa University Graduate School of Medical Science for their unflagging support to our projects. We apologize to all colleagues whose work we could not cite other than indirectly through other publications, due to limitation of space. This work was supported in part by a Grant-in-aid for Scientific Research from the Japan Society for the Promotion of Science (20591067 and 23591329 to H.K), and a Grant-in-Aid for Promotion of Technological Seeds in Advanced Medicine, Hyogo College of Medicine (T.Y.).

9. References

[1] The effect of intensive treatment of diabetes on the development and progression of long-term complications in insulin-dependent diabetes mellitus. The Diabetes Control and Complications Trial Research Group. *N Engl J Med.* 1993;329(14):977-986.

[2] Intensive blood-glucose control with sulphonylureas or insulin compared with conventional treatment and risk of complications in patients with type 2 diabetes (UKPDS 33). UK Prospective Diabetes Study (UKPDS) Group. *Lancet.* 1998;352(9131):837-853.

[3] Patel A, MacMahon S, Chalmers J, Neal B, Billot L, Woodward M, Marre M, Cooper M, Glasziou P, Grobbee D, Hamet P, Harrap S, Heller S, Liu L, Mancia G, Mogensen CE, Pan C, Poulter N, Rodgers A, Williams B, Bompoint S, de Galan BE, Joshi R, Travert F. Intensive blood glucose control and vascular outcomes in patients with type 2 diabetes. *N Engl J Med.* 2008;358(24):2560-2572.

[4] Duckworth W, Abraira C, Moritz T, Reda D, Emanuele N, Reaven PD, Zieve FJ, Marks J, Davis SN, Hayward R, Warren SR, Goldman S, McCarren M, Vitek ME, Henderson WG, Huang GD. Glucose control and vascular complications in veterans with type 2 diabetes. *N Engl J Med.* 2009;360(2):129-139.

[5] Gerstein HC, Miller ME, Byington RP, Goff DC, Jr., Bigger JT, Buse JB, Cushman WC, Genuth S, Ismail-Beigi F, Grimm RH, Jr., Probstfield JL, Simons-Morton DG, Friedewald WT. Effects of intensive glucose lowering in type 2 diabetes. *N Engl J Med.* 2008;358(24):2545-2559.

[6] Nathan DM, Cleary PA, Backlund JY, Genuth SM, Lachin JM, Orchard TJ, Raskin P, Zinman B. Intensive diabetes treatment and cardiovascular disease in patients with type 1 diabetes. *N Engl J Med.* 2005;353(25):2643-2653.

[7] Holman RR, Paul SK, Bethel MA, Matthews DR, Neil HA. 10-year follow-up of intensive glucose control in type 2 diabetes. *N Engl J Med.* 2008;359(15):1577-1589.

[8] Gaede P, Lund-Andersen H, Parving HH, Pedersen O. Effect of a multifactorial intervention on mortality in type 2 diabetes. *N Engl J Med.* 2008;358(6):580-591.

[9] Wanner C, Krane V, Marz W, Olschewski M, Mann JF, Ruf G, Ritz E. Atorvastatin in patients with type 2 diabetes mellitus undergoing hemodialysis. *N Engl J Med.* 2005;353(3):238-248.

[10] Fellstrom BC, Jardine AG, Schmieder RE, Holdaas H, Bannister K, Beutler J, Chae DW, Chevaile A, Cobbe SM, Gronhagen-Riska C, De Lima JJ, Lins R, Mayer G, McMahon AW, Parving HH, Remuzzi G, Samuelsson O, Sonkodi S, Sci D, Suleymanlar G, Tsakiris D, Tesar V, Todorov V, Wiecek A, Wuthrich RP, Gottlow M, Johnsson E, Zannad F. Rosuvastatin and cardiovascular events in patients undergoing hemodialysis. *N Engl J Med.* 2009;360(14):1395-1407.

[11] Koyama H, Yamamoto H, Nishizawa Y. RAGE and soluble RAGE: potential therapeutic targets for cardiovascular diseases. *Mol Med.* 2007;13(11-12):625-635.

[12] Yonekura H, Yamamoto Y, Sakurai S, Petrova RG, Abedin MJ, Li H, Yasui K, Takeuchi M, Makita Z, Takasawa S, Okamoto H, Watanabe T, Yamamoto H. Novel splice variants of the receptor for advanced glycation end-products expressed in human vascular endothelial cells and pericytes, and their putative roles in diabetes-induced vascular injury. *Biochem J.* 2003;370(Pt 3):1097-1109.

[13] Raucci A, Cugusi S, Antonelli A, Barabino SM, Monti L, Bierhaus A, Reiss K, Saftig P, Bianchi ME. A soluble form of the receptor for advanced glycation endproducts (RAGE) is produced by proteolytic cleavage of the membrane-bound form by the

sheddase a disintegrin and metalloprotease 10 (ADAM10). *FASEB J.* 2008;22(10):3716-3727.

[14] Zhang L, Bukulin M, Kojro E, Roth A, Metz VV, Fahrenholz F, Nawroth PP, Bierhaus A, Postina R. Receptor for advanced glycation end products is subjected to protein ectodomain shedding by metalloproteinases. *J Biol Chem.* 2008;283(51):35507-35516.

[15] Koyama H, Shoji T, Yokoyama H, Motoyama K, Mori K, Fukumoto S, Emoto M, Tamei H, Matsuki H, Sakurai S, Yamamoto Y, Yonekura H, Watanabe T, Yamamoto H, Nishizawa Y. Plasma level of endogenous secretory RAGE is associated with components of the metabolic syndrome and atherosclerosis. *Arterioscler Thromb Vasc Biol.* 2005;25(12):2587-2593.

[16] Ueno H, Koyama H, Shoji T, Monden M, Fukumoto S, Tanaka S, Otsuka Y, Mima Y, Morioka T, Mori K, Shioi A, Yamamoto H, Inaba M, Nishizawa Y. Receptor for advanced glycation end-products (RAGE) regulation of adiposity and adiponectin is associated with atherogenesis in apoE-deficient mouse. *Atherosclerosis.* 2010;211(2):431-436.

[17] Hotamisligil GS. Inflammation and metabolic disorders. *Nature.* 2006;444(7121):860-867.

[18] Monnier VM, Cerami A. Nonenzymatic browning in vivo: possible process for aging of long-lived proteins. *Science.* 1981;211(4481):491-493.

[19] Dyer DG, Blackledge JA, Thorpe SR, Baynes JW. Formation of pentosidine during nonenzymatic browning of proteins by glucose. Identification of glucose and other carbohydrates as possible precursors of pentosidine in vivo. *J Biol Chem.* 1991;266(18):11654-11660.

[20] Thorpe SR, Baynes JW. Maillard reaction products in tissue proteins: new products and new perspectives. *Amino Acids.* 2003;25(3-4):275-281.

[21] Wolff SP, Dean RT. Glucose autoxidation and protein modification. The potential role of 'autoxidative glycosylation' in diabetes. *Biochem J.* 1987;245(1):243-250.

[22] Harding JJ, Beswick HT. The possible contribution of glucose autoxidation to protein modification of diabetes. *Biochem J.* 1988;249(2):617-618.

[23] O'Brien J, Morrissey PA. Nutritional and toxicological aspects of the Maillard browning reaction in foods. *Crit Rev Food Sci Nutr.* 1989;28(3):211-248.

[24] Vlassara H, Striker G. Glycotoxins in the diet promote diabetes and diabetic complications. *Curr Diab Rep.* 2007;7(3):235-241.

[25] Koschinsky T, He CJ, Mitsuhashi T, Bucala R, Liu C, Buenting C, Heitmann K, Vlassara H. Orally absorbed reactive glycation products (glycotoxins): an environmental risk factor in diabetic nephropathy. *Proc Natl Acad Sci U S A.* 1997;94(12):6474-6479.

[26] He C, Sabol J, Mitsuhashi T, Vlassara H. Dietary glycotoxins: inhibition of reactive products by aminoguanidine facilitates renal clearance and reduces tissue sequestration. *Diabetes.* 1999;48(6):1308-1315.

[27] Uribarri J, Peppa M, Cai W, Goldberg T, Lu M, He C, Vlassara H. Restriction of dietary glycotoxins reduces excessive advanced glycation end products in renal failure patients. *J Am Soc Nephrol.* 2003;14(3):728-731.

[28] Che W, Asahi M, Takahashi M, Kaneto H, Okado A, Higashiyama S, Taniguchi N. Selective induction of heparin-binding epidermal growth factor-like growth factor by methylglyoxal and 3-deoxyglucosone in rat aortic smooth muscle cells. The involvement of reactive oxygen species formation and a possible implication for atherogenesis in diabetes. *J Biol Chem.* 1997;272(29):18453-18459.

[29] Fu MX, Requena JR, Jenkins AJ, Lyons TJ, Baynes JW, Thorpe SR. The advanced glycation end product, Nepsilon-(carboxymethyl)lysine, is a product of both lipid peroxidation and glycoxidation reactions. *J Biol Chem.* 1996;271(17):9982-9986.

[30] Wells-Knecht KJ, Brinkmann E, Wells-Knecht MC, Litchfield JE, Ahmed MU, Reddy S, Zyzak DV, Thorpe SR, Baynes JW. New biomarkers of Maillard reaction damage to proteins. *Nephrol Dial Transplant.* 1996;11 Suppl 5:41-47.

[31] Baynes JW, Thorpe SR. Role of oxidative stress in diabetic complications: a new perspective on an old paradigm. *Diabetes.* 1999;48(1):1-9.

[32] Miyata T, van Ypersele de Strihou C, Kurokawa K, Baynes JW. Alterations in nonenzymatic biochemistry in uremia: origin and significance of "carbonyl stress" in long-term uremic complications. *Kidney Int.* 1999;55(2):389-399.

[33] Singh R, Barden A, Mori T, Beilin L. Advanced glycation end-products: a review. *Diabetologia.* 2001;44(2):129-146.

[34] Kaneto H, Fujii J, Myint T, Miyazawa N, Islam KN, Kawasaki Y, Suzuki K, Nakamura M, Tatsumi H, Yamasaki Y, Taniguchi N. Reducing sugars trigger oxidative modification and apoptosis in pancreatic beta-cells by provoking oxidative stress through the glycation reaction. *Biochem J.* 1996;320 (Pt 3):855-863.

[35] Zhao Z, Zhao C, Zhang XH, Zheng F, Cai W, Vlassara H, Ma ZA. Advanced glycation end products inhibit glucose-stimulated insulin secretion through nitric oxide-dependent inhibition of cytochrome c oxidase and adenosine triphosphate synthesis. *Endocrinology.* 2009;150(6):2569-2576.

[36] Kuniyasu A, Ohgami N, Hayashi S, Miyazaki A, Horiuchi S, Nakayama H. CD36-mediated endocytic uptake of advanced glycation end products (AGE) in mouse 3T3-L1 and human subcutaneous adipocytes. *FEBS Lett.* 2003;537(1-3):85-90.

[37] Miele C, Riboulet A, Maitan MA, Oriente F, Romano C, Formisano P, Giudicelli J, Beguinot F, Van Obberghen E. Human glycated albumin affects glucose metabolism in L6 skeletal muscle cells by impairing insulin-induced insulin receptor substrate (IRS) signaling through a protein kinase C alpha-mediated mechanism. *J Biol Chem.* 2003;278(48):47376-47387.

[38] Brownlee M. Negative consequences of glycation. *Metabolism.* 2000;49(2 Suppl 1):9-13.

[39] Peppa M, He C, Hattori M, McEvoy R, Zheng F, Vlassara H. Fetal or neonatal low-glycotoxin environment prevents autoimmune diabetes in NOD mice. *Diabetes.* 2003;52(6):1441-1448.

[40] Lin RY, Choudhury RP, Cai W, Lu M, Fallon JT, Fisher EA, Vlassara H. Dietary glycotoxins promote diabetic atherosclerosis in apolipoprotein E-deficient mice. *Atherosclerosis.* 2003;168(2):213-220.

[41] Stirban A, Negrean M, Stratmann B, Gawlowski T, Horstmann T, Gotting C, Kleesiek K, Mueller-Roesel M, Koschinsky T, Uribarri J, Vlassara H, Tschoepe D. Benfotiamine prevents macro- and microvascular endothelial dysfunction and oxidative stress following a meal rich in advanced glycation end products in individuals with type 2 diabetes. *Diabetes Care.* 2006;29(9):2064-2071.

[42] Negrean M, Stirban A, Stratmann B, Gawlowski T, Horstmann T, Gotting C, Kleesiek K, Mueller-Roesel M, Koschinsky T, Uribarri J, Vlassara H, Tschoepe D. Effects of low- and high-advanced glycation endproduct meals on macro- and microvascular endothelial function and oxidative stress in patients with type 2 diabetes mellitus. *Am J Clin Nutr.* 2007;85(5):1236-1243.

[43] Uribarri J, Stirban A, Sander D, Cai W, Negrean M, Buenting CE, Koschinsky T, Vlassara H. Single oral challenge by advanced glycation end products acutely impairs endothelial function in diabetic and nondiabetic subjects. *Diabetes Care.* 2007;30(10):2579-2582.

[44] Linden E, Cai W, He JC, Xue C, Li Z, Winston J, Vlassara H, Uribarri J. Endothelial dysfunction in patients with chronic kidney disease results from advanced

glycation end products (AGE)-mediated inhibition of endothelial nitric oxide synthase through RAGE activation. *Clin J Am Soc Nephrol.* 2008;3(3):691-698.

[45] Sherwin RS, Kramer KJ, Tobin JD, Insel PA, Liljenquist JE, Berman M, Andres R. A model of the kinetics of insulin in man. *J Clin Invest.* 1974;53(5):1481-1492.

[46] Yang YJ, Hope ID, Ader M, Bergman RN. Insulin transport across capillaries is rate limiting for insulin action in dogs. *J Clin Invest.* 1989;84(5):1620-1628.

[47] Jansson PA, Fowelin JP, von Schenck HP, Smith UP, Lonnroth PN. Measurement by microdialysis of the insulin concentration in subcutaneous interstitial fluid. Importance of the endothelial barrier for insulin. *Diabetes.* 1993;42(10):1469-1473.

[48] Barrett EJ, Eggleston EM, Inyard AC, Wang H, Li G, Chai W, Liu Z. The vascular actions of insulin control its delivery to muscle and regulate the rate-limiting step in skeletal muscle insulin action. *Diabetologia.* 2009;52(5):752-764.

[49] Kubota T, Kubota N, Kumagai H, Yamaguchi S, Kozono H, Takahashi T, Inoue M, Itoh S, Takamoto I, Sasako T, Kumagai K, Kawai T, Hashimoto S, Kobayashi T, Sato M, Tokuyama K, Nishimura S, Tsunoda M, Ide T, Murakami K, Yamazaki T, Ezaki O, Kawamura K, Masuda H, Moroi M, Sugi K, Oike Y, Shimokawa H, Yanagihara N, Tsutsui M, Terauchi Y, Tobe K, Nagai R, Kamata K, Inoue K, Kodama T, Ueki K, Kadowaki T. Impaired insulin signaling in endothelial cells reduces insulin-induced glucose uptake by skeletal muscle. *Cell Metab.* 2011;13(3):294-307.

[50] Hofmann SM, Dong HJ, Li Z, Cai W, Altomonte J, Thung SN, Zeng F, Fisher EA, Vlassara H. Improved insulin sensitivity is associated with restricted intake of dietary glycoxidation products in the db/db mouse. *Diabetes.* 2002;51(7):2082-2089.

[51] Harcourt BE, Sourris KC, Coughlan MT, Walker KZ, Dougherty SL, Andrikopoulos S, Morley AL, Thallas-Bonke V, Chand V, Penfold SA, de Courten MP, Thomas MC, Kingwell BA, Bierhaus A, Cooper ME, Courten BD, Forbes JM. Targeted reduction of advanced glycation improves renal function in obesity. *Kidney Int.* 2011 in press.

[52] Sandu O, Song K, Cai W, Zheng F, Uribarri J, Vlassara H. Insulin resistance and type 2 diabetes in high-fat-fed mice are linked to high glycotoxin intake. *Diabetes.* 2005;54(8):2314-2319.

[53] Unno Y, Sakai M, Sakamoto Y, Kuniyasu A, Nakayama H, Nagai R, Horiuchi S. Advanced glycation end products-modified proteins and oxidized LDL mediate down-regulation of leptin in mouse adipocytes via CD36. *Biochem Biophys Res Commun.* 2004;325(1):151-156.

[54] Riboulet-Chavey A, Pierron A, Durand I, Murdaca J, Giudicelli J, Van Obberghen E. Methylglyoxal impairs the insulin signaling pathways independently of the formation of intracellular reactive oxygen species. *Diabetes.* 2006;55(5):1289-1299.

[55] Unoki H, Bujo H, Yamagishi S, Takeuchi M, Imaizumi T, Saito Y. Advanced glycation end products attenuate cellular insulin sensitivity by increasing the generation of intracellular reactive oxygen species in adipocytes. *Diabetes Res Clin Pract.* 2007;76(2):236-244.

[56] Hofmann MA, Drury S, Fu C, Qu W, Taguchi A, Lu Y, Avila C, Kambham N, Bierhaus A, Nawroth P, Neurath MF, Slattery T, Beach D, McClary J, Nagashima M, Morser J, Stern D, Schmidt AM. RAGE mediates a novel proinflammatory axis: a central cell surface receptor for S100/calgranulin polypeptides. *Cell.* 1999;97(7):889-901.

[57] Hori O, Brett J, Slattery T, Cao R, Zhang J, Chen JX, Nagashima M, Lundh ER, Vijay S, Nitecki D, Morser J, Stern D, Schmidt AM. The receptor for advanced glycation end products (RAGE) is a cellular binding site for amphoterin. Mediation of neurite outgrowth and co-expression of rage and amphoterin in the developing nervous system. *J Biol Chem.* 1995;270(43):25752-25761.

[58] Taguchi A, Blood DC, del Toro G, Canet A, Lee DC, Qu W, Tanji N, Lu Y, Lalla E, Fu C, Hofmann MA, Kislinger T, Ingram M, Lu A, Tanaka H, Hori O, Ogawa S, Stern DM, Schmidt AM. Blockade of RAGE-amphoterin signalling suppresses tumour growth and metastases. *Nature*. 2000;405(6784):354-360.

[59] Yan SD, Chen X, Fu J, Chen M, Zhu H, Roher A, Slattery T, Zhao L, Nagashima M, Morser J, Migheli A, Nawroth P, Stern D, Schmidt AM. RAGE and amyloid-beta peptide neurotoxicity in Alzheimer's disease. *Nature*. 1996;382(6593):685-691.

[60] Sousa MM, Yan SD, Stern D, Saraiva MJ. Interaction of the receptor for advanced glycation end products (RAGE) with transthyretin triggers nuclear transcription factor kB (NF-kB) activation. *Lab Invest*. 2000;80(7):1101-1110.

[61] Chavakis T, Bierhaus A, Al-Fakhri N, Schneider D, Witte S, Linn T, Nagashima M, Morser J, Arnold B, Preissner KT, Nawroth PP. The pattern recognition receptor (RAGE) is a counterreceptor for leukocyte integrins: a novel pathway for inflammatory cell recruitment. *J Exp Med*. 2003;198(10):1507-1515.

[62] Haimoto H, Hosoda S, Kato K. Differential distribution of immunoreactive S100-alpha and S100-beta proteins in normal nonnervous human tissues. *Lab Invest*. 1987;57(5):489-498.

[63] Kato K, Suzuki F, Ogasawara N. Induction of S100 protein in 3T3-L1 cells during differentiation to adipocytes and its liberating by lipolytic hormones. *Eur J Biochem*. 1988;177(2):461-466.

[64] Braga CW, Martinez D, Wofchuk S, Portela LV, Souza DO. S100B and NSE serum levels in obstructive sleep apnea syndrome. *Sleep Med*. 2006;7(5):431-435.

[65] Steiner J, Schiltz K, Walter M, Wunderlich MT, Keilhoff G, Brisch R, Bielau H, Bernstein HG, Bogerts B, Schroeter ML, Westphal S. S100B serum levels are closely correlated with body mass index: an important caveat in neuropsychiatric research. *Psychoneuroendocrinology*. 2010;35(2):321-324.

[66] Steiner J, Walter M, Guest P, Myint AM, Schiltz K, Panteli B, Brauner M, Bernstein HG, Gos T, Herberth M, Schroeter ML, Schwarz MJ, Westphal S, Bahn S, Bogerts B. Elevated S100B levels in schizophrenia are associated with insulin resistance. *Mol Psychiatry*. 2010;15(1):3-4.

[67] Lappalainen T, Kolehmainen M, Schwab U, Pulkkinen L, de Mello VD, Vaittinen M, Laaksonen DE, Poutanen K, Uusitupa M, Gylling H. Gene expression of FTO in human subcutaneous adipose tissue, peripheral blood mononuclear cells and adipocyte cell line. *J Nutrigenet Nutrigenomics*. 2010;3(1):37-45.

[68] Weisberg SP, McCann D, Desai M, Rosenbaum M, Leibel RL, Ferrante AW, Jr. Obesity is associated with macrophage accumulation in adipose tissue. *J Clin Invest*. 2003;112(12):1796-1808.

[69] Xu H, Barnes GT, Yang Q, Tan G, Yang D, Chou CJ, Sole J, Nichols A, Ross JS, Tartaglia LA, Chen H. Chronic inflammation in fat plays a crucial role in the development of obesity-related insulin resistance. *J Clin Invest*. 2003;112(12):1821-1830.

[70] Shoelson SE, Lee J, Goldfine AB. Inflammation and insulin resistance. *J Clin Invest*. 2006;116(7):1793-1801.

[71] Schmidt AM, Vianna M, Gerlach M, Brett J, Ryan J, Kao J, Esposito C, Hegarty H, Hurley W, Clauss M, et al. Isolation and characterization of two binding proteins for advanced glycosylation end products from bovine lung which are present on the endothelial cell surface. *J Biol Chem*. 1992;267(21):14987-14997.

[72] Neeper M, Schmidt AM, Brett J, Yan SD, Wang F, Pan YC, Elliston K, Stern D, Shaw A. Cloning and expression of a cell surface receptor for advanced glycosylation end products of proteins. *J Biol Chem*. 1992;267(21):14998-15004.

[73] Kislinger T, Fu C, Huber B, Qu W, Taguchi A, Du Yan S, Hofmann M, Yan SF, Pischetsrieder M, Stern D, Schmidt AM. N(epsilon)-(carboxymethyl)lysine adducts of proteins are ligands for receptor for advanced glycation end products that activate cell signaling pathways and modulate gene expression. *J Biol Chem.* 1999;274(44):31740-31749.

[74] Reddy S, Bichler J, Wells-Knecht KJ, Thorpe SR, Baynes JW. N epsilon-(carboxymethyl)lysine is a dominant advanced glycation end product (AGE) antigen in tissue proteins. *Biochemistry.* 1995;34(34):10872-10878.

[75] Xie J, Reverdatto S, Frolov A, Hoffmann R, Burz DS, Shekhtman A. Structural basis for pattern recognition by the receptor for advanced glycation end products (RAGE). *J Biol Chem.* 2008;283(40):27255-27269.

[76] Matsumoto S, Yoshida T, Murata H, Harada S, Fujita N, Nakamura S, Yamamoto Y, Watanabe T, Yonekura H, Yamamoto H, Ohkubo T, Kobayashi Y. Solution structure of the variable-type domain of the receptor for advanced glycation end products: new insight into AGE-RAGE interaction. *Biochemistry.* 2008;47(47):12299-12311.

[77] Krieger M, Stern DM. Series introduction: multiligand receptors and human disease. *J Clin Invest.* 2001;108(5):645-647.

[78] Schmidt AM, Yan SD, Yan SF, Stern DM. The multiligand receptor RAGE as a progression factor amplifying immune and inflammatory responses. *J. Clin. Invest.* 2001;108(7):949-955.

[79] Schmidt AM, Hori O, Chen JX, Li JF, Crandall J, Zhang J, Cao R, Yan SD, Brett J, Stern D. Advanced glycation endproducts interacting with their endothelial receptor induce expression of vascular cell adhesion molecule-1 (VCAM-1) in cultured human endothelial cells and in mice. A potential mechanism for the accelerated vasculopathy of diabetes. *J Clin Invest.* 1995;96(3):1395-1403.

[80] Basta G, Lazzerini G, Massaro M, Simoncini T, Tanganelli P, Fu C, Kislinger T, Stern DM, Schmidt AM, De Caterina R. Advanced Glycation End Products Activate Endothelium Through Signal-Transduction Receptor RAGE: A Mechanism for Amplification of Inflammatory Responses. *Circulation.* 2002;105(7):816-822.

[81] Harja E, Bu DX, Hudson BI, Chang JS, Shen X, Hallam K, Kalea AZ, Lu Y, Rosario RH, Oruganti S, Nikolla Z, Belov D, Lalla E, Ramasamy R, Yan SF, Schmidt AM. Vascular and inflammatory stresses mediate atherosclerosis via RAGE and its ligands in apoE-/- mice. *J Clin Invest.* 2008;118(1):183-194.

[82] Hirosumi J, Tuncman G, Chang L, Gorgun CZ, Uysal KT, Maeda K, Karin M, Hotamisligil GS. A central role for JNK in obesity and insulin resistance. *Nature.* 2002;420(6913):333-336.

[83] Tuncman G, Hirosumi J, Solinas G, Chang L, Karin M, Hotamisligil GS. Functional in vivo interactions between JNK1 and JNK2 isoforms in obesity and insulin resistance. *Proc Natl Acad Sci U S A.* 2006;103(28):10741-10746.

[84] Basta G, Lazzerini G, Del Turco S, Ratto GM, Schmidt AM, De Caterina R. At least 2 distinct pathways generating reactive oxygen species mediate vascular cell adhesion molecule-1 induction by advanced glycation end products. *Arterioscler Thromb Vasc Biol.* 2005;25(7):1401-1407.

[85] Yan SD, Schmidt AM, Anderson GM, Zhang J, Brett J, Zou YS, Pinsky D, Stern D. Enhanced cellular oxidant stress by the interaction of advanced glycation end products with their receptors/binding proteins. *J Biol Chem.* 1994;269(13):9889-9897.

[86] Furukawa S, Fujita T, Shimabukuro M, Iwaki M, Yamada Y, Nakajima Y, Nakayama O, Makishima M, Matsuda M, Shimomura I. Increased oxidative stress in obesity and its impact on metabolic syndrome. *J Clin Invest*. 2004;114(12):1752-1761.

[87] Houstis N, Rosen ED, Lander ES. Reactive oxygen species have a causal role in multiple forms of insulin resistance. *Nature*. 2006;440(7086):944-948.

[88] Rodino-Janeiro BK, Salgado-Somoza A, Teijeira-Fernandez E, Gonzalez-Juanatey JR, Alvarez E, Eiras S. Receptor for advanced glycation end-products expression in subcutaneous adipose tissue is related to coronary artery disease. *Eur J Endocrinol*. 2011;164(4):529-537.

[89] Orlova VV, Choi EY, Xie C, Chavakis E, Bierhaus A, Ihanus E, Ballantyne CM, Gahmberg CG, Bianchi ME, Nawroth PP, Chavakis T. A novel pathway of HMGB1-mediated inflammatory cell recruitment that requires Mac-1-integrin. *Embo J*. 2007;26(4):1129-1139.

[90] Schmidt AM, Yan SD, Brett J, Mora R, Nowygrod R, Stern D. Regulation of human mononuclear phagocyte migration by cell surface-binding proteins for advanced glycation end products. *J Clin Invest*. 1993;91(5):2155-2168.

[91] Strissel KJ, Stancheva Z, Miyoshi H, Perfield JW, 2nd, DeFuria J, Jick Z, Greenberg AS, Obin MS. Adipocyte death, adipose tissue remodeling, and obesity complications. *Diabetes*. 2007;56(12):2910-2918.

[92] Uusitupa MI, Niskanen LK, Siitonen O, Voutilainen E, Pyorala K. 5-year incidence of atherosclerotic vascular disease in relation to general risk factors, insulin level, and abnormalities in lipoprotein composition in non-insulin-dependent diabetic and nondiabetic subjects. *Circulation*. 1990;82(1):27-36.

[93] Steinberg HO, Chaker H, Leaming R, Johnson A, Brechtel G, Baron AD. Obesity/insulin resistance is associated with endothelial dysfunction. Implications for the syndrome of insulin resistance. *J Clin Invest*. 1996;97(11):2601-2610.

[94] Gao X, Zhang H, Schmidt AM, Zhang C. AGE/RAGE Produces Endothelial Dysfunction in Coronary Arterioles in Type II Diabetic Mice. *Am J Physiol Heart Circ Physiol*. 2008; 295(2): H491-498.

[95] Malherbe P, Richards JG, Gaillard H, Thompson A, Diener C, Schuler A, Huber G. cDNA cloning of a novel secreted isoform of the human receptor for advanced glycation end products and characterization of cells co-expressing cell-surface scavenger receptors and Swedish mutant amyloid precursor protein. *Brain Res Mol Brain Res*. 1999; 71(2):159-170.

[96] Schlueter C, Hauke S, Flohr AM, Rogalla P, Bullerdiek J. Tissue-specific expression patterns of the RAGE receptor and its soluble forms--a result of regulated alternative splicing? *Biochim Biophys Acta*. 2003;1630(1):1-6.

[97] Park IH, Yeon SI, Youn JH, Choi JE, Sasaki N, Choi IH, Shin JS. Expression of a novel secreted splice variant of the receptor for advanced glycation end products (RAGE) in human brain astrocytes and peripheral blood mononuclear cells. *Mol Immunol*. 2004;40(16):1203-1211.

[98] Ding Q, Keller JN. Splice variants of the receptor for advanced glycosylation end products (RAGE) in human brain. *Neurosci Lett*. 2005;373(1):67-72.

[99] Bierhaus A, Humpert PM, Morcos M, Wendt T, Chavakis T, Arnold B, Stern DM, Nawroth PP. Understanding RAGE, the receptor for advanced glycation end products. *J Mol Med*. 2005;83(11):876-886.

[100] Ohe K, Watanabe T, Harada S, Munesue S, Yamamoto Y, Yonekura H, Yamamoto H. Regulation of alternative splicing of the receptor for advanced glycation

endproducts (RAGE) through G-rich cis-elements and heterogenous nuclear ribonucleoprotein H. *J Biochem.* 2010;147(5):651-659.

[101] Shoji T, Koyama H, Morioka T, Tanaka S, Kizu A, Motoyama K, Mori K, Fukumoto S, Shioi A, Shimogaito N, Takeuchi M, Yamamoto Y, Yonekura H, Yamamoto H, Nishizawa Y. Receptor for advanced glycation end products is involved in impaired angiogenic response in diabetes. *Diabetes.* 2006;55(8):2245-2255.

[102] Hudson BI, Harja E, Moser B, Schmidt AM. Soluble levels of receptor for advanced glycation endproducts (sRAGE) and coronary artery disease: the next C-reactive protein? *Arterioscler Thromb Vasc Biol.* 2005;25(5):879-882.

[103] Hanford LE, Enghild JJ, Valnickova Z, Petersen SV, Schaefer LM, Schaefer TM, Reinhart TA, Oury TD. Purification and characterization of mouse soluble receptor for advanced glycation end products (sRAGE). *J Biol Chem.* 2004;279(48):50019-50024.

[104] Gaens KH, Ferreira I, van der Kallen CJ, van Greevenbroek MM, Blaak EE, Feskens EJ, Dekker JM, Nijpels G, Heine RJ, t Hart LM, de Groot PG, Stehouwer CD, Schalkwijk CG. Association of polymorphism in the receptor for advanced glycation end products (RAGE) gene with circulating RAGE levels. *J Clin Endocrinol Metab.* 2009;94(12):5174-5180.

[105] Pullerits R, Brisslert M, Jonsson IM, Tarkowski A. Soluble receptor for advanced glycation end products triggers a proinflammatory cytokine cascade via beta2 integrin Mac-1. *Arthritis Rheum.* 2006;54(12):3898-3907.

[106] Falcone C, Emanuele E, D'Angelo A, Buzzi MP, Belvito C, Cuccia M, Geroldi D. Plasma levels of soluble receptor for advanced glycation end products and coronary artery disease in nondiabetic men. *Arterioscler Thromb Vasc Biol.* 2005;25(5):1032-1037.

[107] Emanuele E, D'Angelo A, Tomaino C, Binetti G, Ghidoni R, Politi P, Bernardi L, Maletta R, Bruni AC, Geroldi D. Circulating levels of soluble receptor for advanced glycation end products in Alzheimer disease and vascular dementia. *Arch Neurol.* 2005;62(11):1734-1736.

[108] Sakurai S, Yamamoto Y, Tamei H, Matsuki H, Obata K, Hui L, Miura J, Osawa M, Uchigata Y, Iwamoto Y, Watanabe T, Yonekura H, Yamamoto H. Development of an ELISA for esRAGE and its application to type 1 diabetic patients. *Diabetes Res Clin Pract.* 2006;73(2):158-165.

[109] Humpert PM, Djuric Z, Kopf S, Rudofsky G, Morcos M, Nawroth PP, Bierhaus A. Soluble RAGE but not endogenous secretory RAGE is associated with albuminuria in patients with type 2 diabetes. *Cardiovasc Diabetol.* 2007;6:9.

[110] Katakami N, Matsuhisa M, Kaneto H, Matsuoka TA, Sakamoto K, Nakatani Y, Ohtoshi K, Hayaishi-Okano R, Kosugi K, Hori M, Yamasaki Y. Decreased endogenous secretory advanced glycation end product receptor in type 1 diabetic patients: its possible association with diabetic vascular complications. *Diabetes Care.* 2005;28(11):2716-2721.

[111] Katakami N, Matsuhisa M, Kaneto H, Yamasaki Y. Serum endogenous secretory RAGE levels are inversely associated with carotid IMT in type 2 diabetic patients. *Atherosclerosis.* 2007;190(1):22-23.

[112] Katakami N, Matsuhisa M, Kaneto H, Matsuoka TA, Sakamoto K, Yasuda T, Umayahara Y, Kosugi K, Yamasaki Y. Serum endogenous secretory RAGE level is an independent risk factor for the progression of carotid atherosclerosis in type 1 diabetes. *Atherosclerosis.* 2009;204(1):288-292.

[113] Lu L, Pu LJ, Zhang Q, Wang LJ, Kang S, Zhang RY, Chen QJ, Wang JG, De Caterina R, Shen WF. Increased glycated albumin and decreased esRAGE levels are related to angiographic severity and extent of coronary artery disease in patients with type 2 diabetes. *Atherosclerosis.* 2009;206(2):540-545.

[114] Koyama H, Shoji T, Fukumoto S, Shinohara K, Emoto M, Mori K, Tahara H, Ishimura E, Kakiya R, Tabata T, Yamamoto H, Nishizawa Y. Low circulating endogenous secretory receptor for AGEs predicts cardiovascular mortality in patients with end-stage renal disease. *Arterioscler Thromb Vasc Biol.* 2007;27(1):147-153.

[115] Yamagishi S, Adachi H, Nakamura K, Matsui T, Jinnouchi Y, Takenaka K, Takeuchi M, Enomoto M, Furuki K, Hino A, Shigeto Y, Imaizumi T. Positive association between serum levels of advanced glycation end products and the soluble form of receptor for advanced glycation end products in nondiabetic subjects. *Metabolism.* 2006;55(9):1227-1231.

[116] Miura J, Yamamoto Y, Osawa M, Watanabe T, Yonekura H, Uchigata Y, Yamamoto H, Iwamoto Y. Endogenous secretory receptor for advanced glycation endproducts levels are correlated with serum pentosidine and CML in patients with type 1 diabetes. *Arterioscler Thromb Vasc Biol.* 2007;27(1):253-254.

[117] Geroldi D, Falcone C, Emanuele E, D'Angelo A, Calcagnino M, Buzzi MP, Scioli GA, Fogari R. Decreased plasma levels of soluble receptor for advanced glycation end-products in patients with essential hypertension. *J Hypertens.* 2005;23(9):1725-1729.

[118] Yilmaz Y, Ulukaya E, Gul OO, Arabul M, Gul CB, Atug O, Oral AY, Aker S, Dolar E. Decreased plasma levels of soluble receptor for advanced glycation endproducts (sRAGE) in patients with nonalcoholic fatty liver disease. *Clin Biochem.* 2009;42(9):802-807.

[119] D'Adamo E, Giannini C, Chiavaroli V, de Giorgis T, Verrotti A, Chiarelli F, Mohn A. What is the significance of soluble and endogenous secretory receptor for advanced glycation end products in liver steatosis in obese prepubertal children? *Antioxid Redox Signal.* 2011;14(6):1167-1172.

[120] Basta G, Sironi AM, Lazzerini G, Del Turco S, Buzzigoli E, Casolaro A, Natali A, Ferrannini E, Gastaldelli A. Circulating soluble receptor for advanced glycation end products is inversely associated with glycemic control and S100A12 protein. *J Clin Endocrinol Metab.* 2006;91(11):4628-4634.

[121] Devangelio E, Santilli F, Formoso G, Ferroni P, Bucciarelli L, Michetti N, Clissa C, Ciabattoni G, Consoli A, Davi G. Soluble RAGE in type 2 diabetes: association with oxidative stress. *Free Radic Biol Med.* 2007;43(4):511-518.

[122] Challier M, Jacqueminet S, Benabdesselam O, Grimaldi A, Beaudeux JL. Increased serum concentrations of soluble receptor for advanced glycation endproducts in patients with type 1 diabetes. *Clin Chem.* 2005;51(9):1749-1750.

[123] Tan KC, Shiu SW, Chow WS, Leng L, Bucala R, Betteridge DJ. Association between serum levels of soluble receptor for advanced glycation end products and circulating advanced glycation end products in type 2 diabetes. *Diabetologia.* 2006;49(11):2756-2762.

[124] Nakamura K, Yamagishi SI, Adachi H, Kurita-Nakamura Y, Matsui T, Yoshida T, Sato A, Imaizumi T. Elevation of soluble form of receptor for advanced glycation end products (sRAGE) in diabetic subjects with coronary artery disease. *Diabetes Metab Res Rev.* 2007;23(5): 368-371.

[125] Yamamoto Y, Miura J, Sakurai S, Watanabe T, Yonekura H, Tamei H, Matsuki H, Obata KI, Uchigata Y, Iwamoto Y, Koyama H, Yamamoto H. Assaying soluble

forms of receptor for advanced glycation end products. *Arterioscler Thromb Vasc Biol.* 2007;27(6):e33-e34.

[126] Humpert PM, Kopf S, Djuric Z, Wendt T, Morcos M, Nawroth PP, Bierhaus A. Plasma sRAGE is independently associated with urinary albumin excretion in type 2 diabetes. *Diabetes Care.* 2006;29(5):1111-1113.

[127] Hudson BI, Moon YP, Kalea AZ, Khatri M, Marquez C, Schmidt AM, Paik MC, Yoshita M, Sacco RL, Decarli C, Wright CB, Elkind MS. Association of serum soluble Receptor for Advanced Glycation End-products with subclinical cerebrovascular disease: The Northern Manhattan Study (NOMAS). *Atherosclerosis.* 2011 in press.

[128] Hou FF, Ren H, Owen WF, Jr, Guo ZJ, Chen PY, Schmidt AM, Miyata T, Zhang X. Enhanced expression of receptor for advanced glycation end products in chronic kidney disease. *J Am Soc Nephrol.* 2004;15(7):1889-1896.

[129] Kalousova M, Hodkova M, Kazderova M, Fialova J, Tesar V, Dusilova-Sulkova S, Zima T. Soluble receptor for advanced glycation end products in patients with decreased renal function. *Am J Kidney Dis.* 2006;47(3):406-411.

[130] Nin JW, Jorsal A, Ferreira I, Schalkwijk CG, Prins MH, Parving HH, Tarnow L, Rossing P, Stehouwer CD. Higher plasma soluble Receptor for Advanced Glycation End Products (sRAGE) levels are associated with incident cardiovascular disease and all-cause mortality in type 1 diabetes: a 12-year follow-up study. *Diabetes.* 2010;59(8):2027-2032.

[131] Kalousova M, Bartosova K, Zima T, Skibova J, Teplan V, Viklicky O. Pregnancy-associated plasma protein a and soluble receptor for advanced glycation end products after kidney transplantation. *Kidney Blood Press Res.* 2007;30(1):31-37.

[132] Wendt TM, Tanji N, Guo J, Kislinger TR, Qu W, Lu Y, Bucciarelli LG, Rong LL, Moser B, Markowitz GS, Stein G, Bierhaus A, Liliensiek B, Arnold B, Nawroth PP, Stern DM, D'Agati VD, Schmidt AM. RAGE drives the development of glomerulosclerosis and implicates podocyte activation in the pathogenesis of diabetic nephropathy. *Am J Pathol.* 2003;162(4):1123-1137.

[133] Bierhaus A, Haslbeck KM, Humpert PM, Liliensiek B, Dehmer T, Morcos M, Sayed AA, Andrassy M, Schiekofer S, Schneider JG, Schulz JB, Heuss D, Neundorfer B, Dierl S, Huber J, Tritschler H, Schmidt AM, Schwaninger M, Haering HU, Schleicher E, Kasper M, Stern DM, Arnold B, Nawroth PP. Loss of pain perception in diabetes is dependent on a receptor of the immunoglobulin superfamily. *J Clin Invest.* 2004;114(12):1741-1751.

[134] Park L, Raman KG, Lee KJ, Lu Y, Ferran LJ, Jr., Chow WS, Stern D, Schmidt AM. Suppression of accelerated diabetic atherosclerosis by the soluble receptor for advanced glycation endproducts. *Nat Med.* 1998;4(9):1025-1031.

[135] Kislinger T, Tanji N, Wendt T, Qu W, Lu Y, Ferran LJ, Jr., Taguchi A, Olson K, Bucciarelli L, Goova M, Hofmann MA, Cataldegirmen G, D'Agati V, Pischetsrieder M, Stern DM, Schmidt AM. Receptor for advanced glycation end products mediates inflammation and enhanced expression of tissue factor in vasculature of diabetic apolipoprotein E-null mice. *Arterioscler Thromb Vasc Biol.* 2001;21(6):905-910.

[136] Bucciarelli LG, Wendt T, Qu W, Lu Y, Lalla E, Rong LL, Goova MT, Moser B, Kislinger T, Lee DC, Kashyap Y, Stern DM, Schmidt AM. RAGE blockade stabilizes established atherosclerosis in diabetic apolipoprotein E-null mice. *Circulation.* 2002;106(22):2827-2835.

[137] Sakaguchi T, Yan SF, Yan SD, Belov D, Rong LL, Sousa M, Andrassy M, Marso SP, Duda S, Arnold B, Liliensiek B, Nawroth PP, Stern DM, Schmidt AM, Naka Y.

Central role of RAGE-dependent neointimal expansion in arterial restenosis. *J Clin Invest*. 2003;111(7):959-972.

[138] Chen Y, Yan SS, Colgan J, Zhang HP, Luban J, Schmidt AM, Stern D, Herold KC. Blockade of late stages of autoimmune diabetes by inhibition of the receptor for advanced glycation end products. *J Immunol*. 2004;173(2):1399-1405.

[139] Goova MT, Li J, Kislinger T, Qu W, Lu Y, Bucciarelli LG, Nowygrod S, Wolf BM, Caliste X, Yan SF, Stern DM, Schmidt AM. Blockade of receptor for advanced glycation end-products restores effective wound healing in diabetic mice. *Am J Pathol*. 2001;159(2):513-525.

[140] Arancio O, Zhang HP, Chen X, Lin C, Trinchese F, Puzzo D, Liu S, Hegde A, Yan SF, Stern A, Luddy JS, Lue LF, Walker DG, Roher A, Buttini M, Mucke L, Li W, Schmidt AM, Kindy M, Hyslop PA, Stern DM, Du Yan SS. RAGE potentiates Abeta-induced perturbation of neuronal function in transgenic mice. *Embo J*. 2004;23(20):4096-4105.

[141] Cheng C, Tsuneyama K, Kominami R, Shinohara H, Sakurai S, Yonekura H, Watanabe T, Takano Y, Yamamoto H, Yamamoto Y. Expression profiling of endogenous secretory receptor for advanced glycation end products in human organs. *Mod Pathol*. 2005;18(10):1385-1396.

[142] Tanaka N, Yonekura H, Yamagishi S, Fujimori H, Yamamoto Y, Yamamoto H. The receptor for advanced glycation end products is induced by the glycation products themselves and tumor necrosis factor-alpha through nuclear factor-kappa B, and by 17beta-estradiol through Sp-1 in human vascular endothelial cells. *J Biol Chem*. 2000;275(33):25781-25790.

[143] Coughlan MT, Thallas-Bonke V, Pete J, Long DM, Gasser A, Tong DC, Arnstein M, Thorpe SR, Cooper ME, Forbes JM. Combination therapy with the advanced glycation end product cross-link breaker, alagebrium, and angiotensin converting enzyme inhibitors in diabetes: synergy or redundancy? *Endocrinology*. 2007;148(2):886-895.

[144] Zhong Y, Li SH, Liu SM, Szmitko PE, He XQ, Fedak PW, Verma S. C-Reactive protein upregulates receptor for advanced glycation end products expression in human endothelial cells. *Hypertension*. 2006;48(3):504-511.

[145] Geroldi D, Falcone C, Minoretti P, Emanuele E, Arra M, D'Angelo A. High levels of soluble receptor for advanced glycation end products may be a marker of extreme longevity in humans. *J Am Geriatr Soc*. 2006;54(7):1149-1150.

[146] Forbes JM, Thorpe SR, Thallas-Bonke V, Pete J, Thomas MC, Deemer ER, Bassal S, El-Osta A, Long DM, Panagiotopoulos S, Jerums G, Osicka TM, Cooper ME. Modulation of soluble receptor for advanced glycation end products by angiotensin-converting enzyme-1 inhibition in diabetic nephropathy. *J Am Soc Nephrol*. 2005;16(8):2363-2372.

[147] Marx N, Walcher D, Ivanova N, Rautzenberg K, Jung A, Friedl R, Hombach V, de Caterina R, Basta G, Wautier MP, Wautiers JL. Thiazolidinediones reduce endothelial expression of receptors for advanced glycation end products. *Diabetes*. 2004;53(10):2662-2668.

[148] Okamoto T, Yamagishi S, Inagaki Y, Amano S, Koga K, Abe R, Takeuchi M, Ohno S, Yoshimura A, Makita Z. Angiogenesis induced by advanced glycation end products and its prevention by cerivastatin. *Faseb J*. 2002;16(14):1928-1930.

[149] Cuccurullo C, Iezzi A, Fazia ML, De Cesare D, Di Francesco A, Muraro R, Bei R, Ucchino S, Spigonardo F, Chiarelli F, Schmidt AM, Cuccurullo F, Mezzetti A,

Cipollone F. Suppression of RAGE as a basis of simvastatin-dependent plaque stabilization in type 2 diabetes. *Arterioscler Thromb Vasc Biol.* 2006;26(12):2716-2723.

[150] Tam HL, Shiu SW, Wong Y, Chow WS, Betteridge DJ, Tan KC. Effects of atorvastatin on serum soluble receptors for advanced glycation end-products in type 2 diabetes. *Atherosclerosis.* 2010;209(1):173-177.

[151] Tan KC, Chow WS, Tso AW, Xu A, Tse HF, Hoo RL, Betteridge DJ, Lam KS. Thiazolidinedione increases serum soluble receptor for advanced glycation end-products in type 2 diabetes. *Diabetologia.* 2007;50(9):1819-1825.

[152] Oz Gul O, Tuncel E, Yilmaz Y, Ulukaya E, Gul CB, Kiyici S, Oral AY, Guclu M, Ersoy C, Imamoglu S. Comparative effects of pioglitazone and rosiglitazone on plasma levels of soluble receptor for advanced glycation end products in type 2 diabetes mellitus patients. *Metabolism.* 2010;59(1):64-69.

[153] Yan XX, Lu L, Peng WH, Wang LJ, Zhang Q, Zhang RY, Chen QJ, Shen WF. Increased serum HMGB1 level is associated with coronary artery disease in nondiabetic and type 2 diabetic patients. *Atherosclerosis.* 2009;205(2):544-548.

[154] Mahajan N, Malik N, Bahl A, Sharma Y, Dhawan V. Correlation among soluble markers and severity of disease in non-diabetic subjects with pre-mature coronary artery disease. *Mol Cell Biochem.* 2009;330(1-2):201-209.

[155] Basta G, Corciu AI, Vianello A, Del Turco S, Foffa I, Navarra T, Chiappino D, Berti S, Mazzone A. Circulating soluble receptor for advanced glycation end-product levels are decreased in patients with calcific aortic valve stenosis. *Atherosclerosis.* 2011 in press;210(2):614-618.

[156] Basta G, Leonardis D, Mallamaci F, Cutrupi S, Pizzini P, Gaetano L, Tripepi R, Tripepi G, De Caterina R, Zoccali C. Circulating soluble receptor of advanced glycation end product inversely correlates with atherosclerosis in patients with chronic kidney disease. *Kidney Int.* 2010;77(3):225-231.

[157] Yokota C, Minematsu K, Tomii Y, Naganuma M, Ito A, Nagasawa H, Yamaguchi T. Low levels of plasma soluble receptor for advanced glycation end products are associated with severe leukoaraiosis in acute stroke patients. *J Neurol Sci.* 2009;287(1-2):41-44.

[158] Park HY, Yun KH, Park DS. Levels of Soluble Receptor for Advanced Glycation End Products in Acute Ischemic Stroke without a Source of Cardioembolism. *J Clin Neurol.* 2009;5(3):126-132.

[159] Yao L, Li K, Zhang L, Yao S, Piao Z, Song L. Influence of the Pro12Ala polymorphism of PPAR-gamma on age at onset and sRAGE levels in Alzheimer's disease. *Brain Res.* 2009;1291:133-139.

[160] Chiang KH, Huang PH, Huang SS, Wu TC, Chen JW, Lin SJ. Plasma levels of soluble receptor for advanced glycation end products are associated with endothelial function and predict cardiovascular events in nondiabetic patients. *Coron Artery Dis.* 2009;20(4):267-273.

[161] Kalousova M, Jachymova M, Mestek O, Hodkova M, Kazderova M, Tesar V, Zima T. Receptor for advanced glycation end products--soluble form and gene polymorphisms in chronic haemodialysis patients. *Nephrol Dial Transplant.* 2007;22(7):2020-2026.

[162] Semba RD, Ferrucci L, Fink JC, Sun K, Beck J, Dalal M, Guralnik JM, Fried LP. Advanced glycation end products and their circulating receptors and level of kidney function in older community-dwelling women. *Am J Kidney Dis.* 2009;53(1):51-58.

[163] Nakamura K, Yamagishi S, Adachi H, Kurita-Nakamura Y, Matsui T, Yoshida T, Imaizumi T. Serum levels of sRAGE, the soluble form of receptor for advanced glycation end products, are associated with inflammatory markers in patients with type 2 diabetes. *Mol Med*. 2007;13(3-4):185-189.
[164] Nakamura K, Yamagishi SI, Adachi H, Matsui T, Kurita-Nakamura Y, Takeuchi M, Inoue H, Imaizumi T. Circulating advanced glycation end products (AGEs) and soluble form of receptor for AGEs (sRAGE) are independent determinants of serum monocyte chemoattractant protein-1 (MCP-1) levels in patients with type 2 diabetes. *Diabetes Metab Res Rev*. 2007.
[165] Nozaki I, Watanabe T, Kawaguchi M, Akatsu H, Tsuneyama K, Yamamoto Y, Ohe K, Yonekura H, Yamada M, Yamamoto H. Reduced expression of endogenous secretory receptor for advanced glycation endproducts in hippocampal neurons of Alzheimer's disease brains. *Arch Histol Cytol*. 2007;70(5):279-290.

Mitochondria Function in Diabetes – From Health to Pathology – New Perspectives for Treatment of Diabetes-Driven Disorders

Magdalena Labieniec-Watala[1*], Karolina Siewiera[1],
Slawomir Gierszewski[1] and Cezary Watala[2]
[1]*University of Lodz, Department of Thermobiology,*
[2]*Medical University of Lodz, Department of Haemostasis and Haemostatic Disorders,*
Poland

1. Introduction

A few words about mitochondria ….

Mitochondria are active intracellular structures that collide, divide, and fuse with other mitochondria. Mitochondria exist as branched-chain reticulum networks, but can also exist as punctuated structures. Their distribution within the cells is quite diverse, is regulated by the interactions with the cytoskeleton and maintained by the balance between mitochondrial fusion and fission.

Mitochondria play crucial physiological functions that underlie the distortions of fragile balance between health and disease. In recent years, the role of mitochondria has gained much interest in the field of diabetic pathology since mitochondrial abnormalities were found in insulin resistance and in both types of diabetes.

A few words about hyperglycaemia and diabetes….

Hyperglycaemia, resulting from uncontrolled regulation of glucose metabolism, is widely recognized as the causal link between diabetes and diabetic complications. Diabetes mellitus has been classified into two forms. Type 1 diabetes, which accounts for about 10% of all cases of diabetes, is caused by autoimmune destruction of pancreatic β-cells that implies insulin deficiency. Type 2 diabetes, the more prevalent form of diabetes, is considered a heterogeneous disease due to the multiplicity of factors that cause the observed phenotype. It results from the combination of insulin resistance and/or a β-cell secretory defects. The explosive increase in the prevalence of type 2 diabetes is predicted in the nearest future. It is considered that about 220 millions people all over the world may suffer from the condition, and an equal number is thought to be "prediabetic", having early symptoms and not yet full manifestation of the disease.

A few words about impact of hyperglycaemia on mitochondria …..

Hyperglycaemia has been indicated as one of the main causes of altered mitochondrial function in diabetic individuals (animals and humans). Rather huge variation in the extent of damage has been observed and attributed to both the type and duration of diabetes studied. General consensus points out to the formation of glycation products as a crucial mechanism, by which hyperglycaemia affects mitochondrial and cellular function in diabetes.

Hyperglycaemia elicits an increased ROS production, presumably to the major extent originating from mitochondrial respiratory chain. ROS play a central role in mediating various metabolic defects associated with a diabetic state. Therefore, the inhibition of ROS production and/or enhancement of ROS scavenging might prove to be beneficial therapies. Alterations in metabolic regulators and glucose-stimulated insulin secretion are also associated with mitochondrial dysfunction in diabetes.

Impairments in mitochondrial function are intrinsically related to diabetes. The prevailing hypothesis is that hyperglycaemia-induced increase in electron transfer donors (NADH and $FADH_2$) may increase electron flux through the mitochondrial electron transport chain. Consequently, the ATP/ADP ratio and hyperpolarisation of the mitochondrial membrane (electrochemical potential difference) also become increased. This high electrochemical potential difference generated by the proton gradient leads to partial inhibition of the electron transport in the complex III, resulting in the augmented electron flow towards coenzyme Q. In turn, this drives partial reduction of O_2 to generate the free radical anion superoxide. It is accelerated reduction of coenzyme Q and generation of ROS that are believed to constitute the fundamental source for mitochondrial dysfunction that plays a critical role in diabetes-related metabolic disorders and tissue histopathology.

A few words about oxidative stress.....

Oxidative stress has been implicated as a major contributor to both the onset and the progression of diabetes and its associated complications. Some of the consequences of an oxidative environment may be the development of insulin resistance, β-cell dysfunction, impaired glucose tolerance, and mitochondrial dysfunction, which can ultimately lead to the diabetic disease state. Experimental and clinical data suggest an inverse association between insulin sensitivity and ROS levels. Oxidative stress can arise from a number of different sources, like the disease state or lifestyle, including episodes of ketosis, sleep restriction, and excessive nutrient intake. Oxidative stress can be reduced by controlling calorie intake, hyperglycaemia and mitochondrial metabolism.

It is now established that 90% of intracellular ROS are generated by mitochondria. The mitochondrial respiratory chain is the principal source of cellular oxygen radicals (ROS), such as superoxide anion radicals and hydroxyl radicals. The primary factor governing mitochondrial ROS generation is the redox state of the respiratory chain. If the membrane potential across the inner mitochondrial membrane rises above a certain threshold value, a massive stimulation of ROS generation occurs. Electrons leak mainly from the complexes I and III of the electron transport chain (ETC) and thereby generate incompletely reduced forms of oxygen. The rise in a membrane potential may occur as a consequence of augmented delivery of electrons to the respiratory chain, which results from either increased glucose or fatty acid oxidation or as a result of altered ETC stoichiometry. Consequently, an increased reverse electron flow occurs. Also, there is an evidence that increased cytosolic generation of ROS might precipitate increased mitochondrial ROS. The balance between the genesis of physiological mitochondrial ROS and antioxidant defenses may thus become disturbed, causing numerous pathological events, and finally leading to cell death.

A few words about an interesting association between mitochondrial ATP production and ATP deficiency in pathology....

Mitochondria are the primary source of ATP production in every cell. Therefore, disruption of mitochondrial respiratory function is regarded as a key event in the development of pathologic complications due to ATP depletion in different tissues, like for instance in heart tissue in diabetic patients. However, no general consensus has been raised about the occurrence of mitochondrial defects under diabetic conditions. Evidence associating

Mitochondria Function in Diabetes – From Health to Pathology – New Perspectives for Treatment of Diabetes-Driven Disorders

125

diabetes with impaired mitochondrial respiratory function in the liver, heart and kidney of diabetic animals dates back more than 45 years. Despite this long history of research, we have still no comprehensive knowledge on the nature and extent of mitochondrial dysfunction in diabetics, as well as about the mechanisms linking this secondary metabolic abnormality with the primary metabolic defect in insulin and hyperglycaemia. The relationship between mitochondrial dysfunction and diabetic pathology has also not yet been defined and elucidated.

In summary…..

In this chapter we discuss how to possibly modulate the "vicious circle" established between mitochondria, oxidative stress and hyperglycaemia. The potential application of some existing and some new agents possessing promising anti-glycation properties to reduce glycation phenomenon and to increase the antioxidant defense system by targeting mitochondria is discussed. Moreover, this chapter outlines various mechanisms present in mitochondria that may lead to the development of diabetes. Intervention and therapy that alter or disrupt these mechanisms may serve to reduce the risk of development of this pathology.

2. Impact of hyperglycaemia on cellular biochemistry/metabolism – overview of the recent achievements in the field

Glycation and oxidative stress are two important processes known to play a key role in the etiopathology of complications in numerous disease processes. Oxidative stress, either via increasing reactive oxygen species (ROS), or by depleting the antioxidants, may modulate the genesis of glycated proteins *in vitro*, as well as *in vivo*.

2.1 Hyperglycaemia – the basic knowledge on Louis Maillard's discovery

Glycation (non-enzymatic N-glycosylation) is an endogenous process that contributes to the post-translational modification of proteins. It is slow under normal physiological conditions, giving rise to the presence of lysine- and arginine-derived glycation adducts in cellular and extracellular proteins. Inside cells, the impact of glycation is countered by high turnover and short half-life of numerous cellular proteins. In long-lived extracellular proteins, however, glycation adducts accumulate with age (Sell et al., 1996). Then, some of these adducts may be removed by enzymatic repair mechanisms, whilst all are removed by degradation of the glycated proteins. Degradation of extracellular glycated proteins requires specific recognition by receptors, internalisation and proteolytic processing. There are specific receptors, AGE receptors, which fulfill this role (Thornalley, 1998).

The Maillard reaction is named after Louis Maillard, who discovered over 80 years ago that some amines and reducing carbohydrates react to produce brown pigments (Ellis, 1959). The Maillard reaction proceeds via three major stages (early, advanced and final stage) and is dependent upon factors such as pH, time, temperature, as well as type and concentrations of reactants. Maillard reactions occur both *in vivo* and *in vitro*, and are associated with the chronic complications of diabetes, aging and age-related diseases (Edeas et al., 2010). The first step of this reaction typically involves the nucleophilic addition of a reducing sugar to a primary amine group (e.g. as found on a lysine or at the N-terminus of a protein). In this stage a reversible Schiff base is formed, which can undergo a slow irreversible rearrangement to form more stable Amadori product that accumulates over time (Fig. 1). The total amount of such accumulated products is known to be dependent on the type of sugar that is causing the glycation, the incubation time and sugar concentration, as well as the type protein that is being modified (Barnaby et al., 2011).

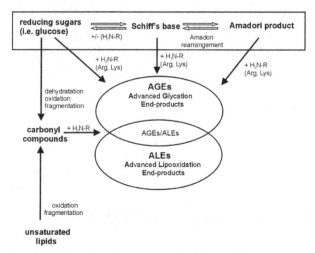

Fig. 1. General scheme of Maillard pathways in diabetic organism.

The first product of Maillard reaction is a simple glycosylamine, which readily undergoes the Amadori rearrangement to produce 1-amino-1-deoxy-2-ketoses. The large body of literature on these reactions is due to the multitude of possible reaction pathways and products, including fragmentations of the carbohydrates and formation of aromatic compounds from cyclisation/dehydration processes (Hodge, 1955). Reducing disaccharides also undergo this reaction, and it is a well-documented process for the degradation of lactose during the heating of milk. Reducing carbohydrates such as glucose, maltose, and lactose are tautomers and are in equilibrium with their more reactive aldehyde forms; nonreducing carbohydrates, such as mannitol, sucrose and trehalose, are not subject to Maillard reactions. Although early scientists believed that only primary aromatic amines were capable to become glycated, subsequent research has shown that nearly all primary and secondary amines, both aromatic and aliphatic, are capable of this reaction (Wirth et al., 1998).

Fragmentation of glucose adducts in early glycation processes establishes many parallel glycation pathways that lead to the subsequent formation of the so-called Advanced Glycation End-products (AGEs). Analogous oxidation and dehydration reactions have been found in glycation by other hexose and pentose derivatives. Glyoxal, methylglyoxal and 3-deoxyglucosone (3-DG)-derived AGEs may be present in proteins glycated by glucose. Methylglyoxal-derived AGEs are common to proteins modified by glucose and by the authentic α-oxoaldehyde. Indeed, similar binding to AGE receptors has been found for these proteins. The formation of α-oxoaldehydes from monosaccharides, Schiff's bases and fructosamines suggests that AGEs may be formed at all stages of glycation (Westwood et al., 1999). AGEs alter structure and functions of proteins. It has been shown that the formation of AGEs *in vivo* contributes to several pathophysiological impairments associated with aging and diabetes mellitus, such as chronic renal insufficiency, Alzheimer's disease, nephropathy, neuropathy and cataract (Ravelojaona et al., 2007).

2.2 Impact of AGEs on human organism – undesirable effects on our health

Glycation of amino-groups on small or large cell constituents induces a number of undesirable effects in a plethora of age-related pathologies, overall referred to as glycation-induced health hazards and including cardiovascular diseases, kidney insufficiencies, retinopathy, or effects of AGEs on embryonic development, as observed in diabetes-related gravidities. As such reactions proceed with a speed proportional to the concentrations of the interacting substances, hyperglycaemia is an important factor for its acceleration (Urios et al., 2007).

It is very important to remember that Maillard products derive also from "ready made" ingested food. Table 1 shows the contents of Maillard products in some foods (expressed as N^{ε}-carboxylysine), as selected from data published by Goldberg et al. (2004).

Name of the selected food	AGE content [U/g]*
Bread Whole wheat, crust, toasted	1.39
Corn flakes	2.32
Peanut Butter Chocolate	32
Popcorn, microwave	336
Butter	265
Fruits: Apple	127
Apple baked	445
Banana	87
Vegetables: Broccoli, carrots, celery	2.26
Carrots, canned	103
Pepper, mushrooms	2.66
Tomato, raw	234
Liquids: Milk, whole	48
Formula, infant	486
Human milk, fresh	52
Apple juice	20
Orange juice, carton	56
Beverages: Coffee, instatnt	53
Tea	19
Cola	65
Condiments: Ketchup	103
Mustard	29
Vinegar sauce, white	377
Cheese: Feta	84
Mozzarella	17
Parmesan	169
Hamburger, fast food	54

*AGE denotes N-carboxymethyllysine (CML)-like immunoreactivity, assessed by enzyme-linked immunosorbent assay using monoclonal antibody 4G9.

Table 1. The content of Maillard products in the selected victuals.

The Maillard reaction between reducing sugars and amino acids is a common reaction in foods, which undergo thermal processing. Desired consequences, like the formation of flavor and brown color of some cooked foods, but also the destruction of essential amino acids and the production of anti-nutritive compounds, require to consider the relevant mechanisms for controlling of Maillard reaction intermediates and final products. Processes such as roasting, baking or frying rely on favorable effects of the Maillard reaction, such as color and flavor formation, whereas during drying, pasteurisation and sterilisation the occurrence of the Maillard reaction is unfavorable. Nutritional losses of essential amino acids that are involved in the reaction, as well as the formation of reaction products are among those unwanted effects (Jaeger et al., 2010).

There is a limited number of studies that have been used to investigate the health effects of dietary Maillard neoformed compounds in humans. Some observational studies have been carried out to address the question of absorption, biodistribution and elimination of dietary Maillard Reaction Products (MRP), and to observe the associations between food exposure to MRPs and their *in vivo* levels.

Some reports have shown that in tobacco leaves, which are dried in the presence of sugars, the Maillard reaction cascade leads to a formation of glycated and oxidative derivatives. These compounds are inhaled during the smoking, after that they are absorbed by lungs and conjugated with serum proteins. It was evidenced that total serum AGE level in cigarette smokers is significantly higher in comparison with non-smokers. However, the highest level of AGEs was detected in the arteries and ocular lenses in diabetic smokers (Vlassara & Palace, 2002).

Furthermore, high AGE levels were observed in industrially preprocessed foods from animal products, like frankfurters, bacon, and powdered egg whites, compared with the unprocessed forms. Across all categories, exposure to higher temperature most of all raised the AGE content (for equal food weights). The temperature level appeared to be more critical than the duration. Also, microwaving increased AGE content more rapidly compared with conventional cooking methods (Peppa et al., 2002). Based on the above data, it is well evidenced that dietary glycoxidation products may constitute an important link between the increased consumption of animal fat and meat and the subsequent development of diabetic complications. However, the problem of AGEs' presence in food is well known, and therefore presently scientists call to use diets containing low contents of these compounds undesirable for out health.

Paradoxically, because of the metabolic demands of the brain, the human body has an obligatory requirement for glucose, approaching 200 g/day. The blood glucose concentration is tightly regulated by homeostatic regulatory systems and maintained between 40 mg/dl (2.2 mmol/l) and 180 mg/dl (10.0 mmol/l). Hypoglycaemia below the lower limit may result in coma, seizures, or even death. Hyperglycaemia, exceeding the upper limit, is associated with immediate glycosuria and caloric loss, as well as long-term consequences, like retinopathy, atherosclerosis, renal failure, etc. Under normal physiological conditions hyperglycaemia stimulates insulin secretion, promoting uptake of glucose by muscles and adipose tissue (Chiu & Taylor, 2011).

Nevertheless, several studies suggest that some MRPs present in foods could have beneficial effects on human health. For instance, the melanoidins are brown Maillard polymers, which seem to have functional properties in food products and are also capable of inhibiting growth of a tumour cell line in culture (Marko et al., 2003). In addition, it was also found recently that a selection of foods rich in MRPs could inhibit the oxidation of LDL *in vitro*.

The high diversity of the MRPs formed in the very diverse food matrices makes it impossible to classify all of them as glycotoxins. It is admitted that they have different beneficial or detrimental biological activities. Thus, more well-controlled clinical experiments are needed to establish the role of the ingested MRPs, pure or added to food matrices, following acute or chronic exposures (Tessier & Birlouez-Aragon, 2010).

2.3 The role of Reactive Oxygen Species (ROS) in glycation process

Free radicals in biological materials were discovered less than 60 years ago. Soon thereafter, Denham Harman hypothesized that reactive oxygen radicals may be formed as by-products of enzymatic reactions *in vivo*. In 1956 he described free radicals as a Pandora's box of evils that may account for gross cellular damage, mutagenesis, cancer, and, last but not the least, the degenerative process of biological aging (Harman, 1956). Presently, the list of cell and tissue disorders caused by free radicals is very long and the diseases, such as diabetes and/or impairments in mitochondria functions, also belong to the "victims" of ROS attack. Glycation and oxidative stress are closely linked, and both phenomena coincide in a vicious process referred to as "glycoxidation". In all steps of glycoxidation there is a massive generation of oxygen-free radicals, some of them being common with lipidic peroxidation pathways. Besides, glycated proteins, and especially their advanced adducts, activate membrane receptors, such as RAGE, and induce an intracellular oxidative stress and a pro-inflammatory status. Glycated proteins may modulate functions of cells involved in oxidative metabolism and induce inappropriate responses. Finally, some oxidative products (reactive aldehydes such as methylglyoxal) or lipid peroxidation products (malondialdehyde) may bind to proteins and amplify glycoxidation generated lesions (Hunt et al., 1998).

Recently, oxygen free radicals, antioxidant defences and the cellular redox status have been considered as central players in pathogenesis of diabetes. The role of glycaemic control on the pro-oxidant/antioxidant balance deserves special attention. Metabolic disturbances and oxidative stress seem to be closely related, improved glycaemic control being associated with a lowered pro-oxidant status (Wierusz-Wysocka et al., 1995).

It was also evidenced that there is a relationship between oxidative stress and insulin resistance observed in diabetes. Hyperinsulinaemia increases the concentrations of ROS, which, in turn, may be responsible for the impaired intracellular insulin actions. Amongst ROS, hydrogen peroxide has been shown to contribute to insulin receptor signaling, and may play a key role in the modulation of the signalling transduction pathways regulated by insulin through coupled receptors.

Consequently, the inactivation of hydrogen peroxide by catalase could represent a critical step for the removal of intracellular ROS in insulin-producing cells. On the other side, the inhibition of catalase under conditions of insulin resistance could also represent an adaptive response to maintain the homeostasis of intracellular hydrogen peroxide as an intermediate of the insulin-activated physiological processes. Overall, relationships between ROS and diabetes seem extremely complex (Bonnefont-Rousselot, 2002).

A second source of ROS formation is an excessive production of AGEs, especially due to a hyperglycaemia-induced overproduction of methylglyoxal. AGEs are also able to produce oxygenated free radicals via complex biochemical mechanisms. AGEs have been shown to interact with their specific receptors (RAGE) and thus they induce oxidative stress, enhance vascular cell adhesion molecule type 1 (VCAM-1) expression, and increase endothelial adhesiveness for monocytes. This overproduction of AGEs appears to play a key role in the pathogenesis of diabetic complications. In particular, the accumulation of two AGEs biomarkers, namely carboxymethyllysine and pentosidine, has been related to the severity of diabetic nephropathy and the so-called 'carbonyl stress'. The toxic effects of AGEs result from structural and functional alterations in proteins, especially the cross-linking of proteins, and from their interactions with RAGEs leading to the enhanced formation of oxygen free radicals (Miyata et al., 2001; Singh et al., 2001).

2.4 Diabetes – a frequent disease or an epidemic?

In 1993 the World Health Organisation (WHO) Ad Hoc Diabetes Reporting Group published standardized global estimates for the prevalence of diabetes and impaired glucose tolerance in adults, based on data from 75 communities in 32 countries. These estimates provided, for the first time, comparable information on the prevalence of abnormal glucose tolerance from many populations worldwide. However, they did not meet the needs of those who frequently refer to the WHO diabetes program for information on the number of people with diabetes in a particular country/community, nor did they take account of future trends in the burden of diabetes (King & Rewers, 1993). Therefore, a further study has been undertaken that links data from the global database collected by WHO with demographic estimates and projections issued by the United Nations to estimate the number of people with diabetes in all countries of the world at three points in time, i.e., the years 1995, 2000, and 2025. The results of this study suggest that for the world as a whole, between the years 1995 and 2025, the adult population will increase by 72%, prevalence of diabetes in adults will increase by 35%, and the number of people with diabetes will increase by 122% (Fig. 2). For the developed countries, there will be an 11% increase in the adult population, a 27% increase in the prevalence of adult diabetes, and a 42% increase in the number of people with diabetes. For the developing countries, there will be an 82% increase in the adult population, a 48% increase in the prevalence of adult diabetes, and a 170% increase in the number of people with diabetes (King et al., 1998).

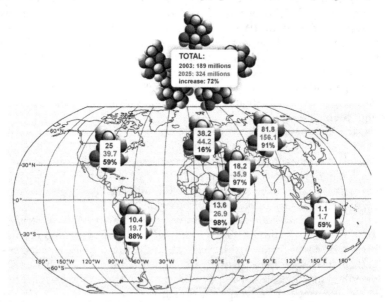

Fig. 2. Global projections for the diabetes epidemic in years 2003-2025.

It is well known that diabetes is one of the most costly and burdensome chronic diseases of our time and is a condition that is increasing in epidemic proportions throughout the world. The prevalence of abnormal glucose tolerance in any population is of public health concern, since diabetes may increase disability burden and health care utilisation.

The relationship between blood glucose concentration in diabetes and the incidence of disease complications was demonstrated in large epidemiological studies. Accurate metabolic control in diabetes is not always feasible, and therefore the issue of molecular mechanisms underlying the damaging effects of hyperglycaemia on body cells and tissues, as well as the possibilities of their pharmacological inhibition, are of the utmost importance in a diabetological practice and anti-diabetic treatment.

The complications resulting from the disease are a significant cause of morbidity and mortality and are associated with the damage or failure of various organs such as eyes, kidneys, and nerves. Although the treatment of diabetes has become increasingly sophisticated, with over a dozen pharmacological agents available to lower blood glucose, a multitude of ancillary supplies and equipment available, and a clear recognition by health care professionals and patients that diabetes is a serious disease, the normalisation of blood glucose for any appreciable period of time is seldom achieved. In addition, in well-controlled so called "intensively" treated patients, serious complications still occur, and the economic and personal burden of diabetes remains (Turner et al., 1999).

Nowadays, diabetes is treated not only as a disease, but as an epidemic. However, as a discipline, diabetes epidemiology is relatively young. The first significant gathering of researches interested in diabetes epidemiology took place just in 1978. Then, in the relatively short span of 2 decades, epidemiology studies have had a profound impact on diabetes research, care and prevention. This explosion of interest and activity in the epidemiology of diabetes should contribute to an effective reduction in the number of patients with this disease.

3. Mitochondria – the relationship between the structure and the function

Mitochondria are multifunction organelles, which play a key role in both the proper functioning of the cell and normal cell death scenario (Kuznetsov & Margreiter, 2009, as cited in McBride et al., 2006). Their main role is the production of adenosine triphosphate (ATP) through metabolic processes involving tricarboxylic acid cycle (TCA) and the electron transport chain (ETC). Most cellular ATP is generated in the process of oxidative phosphorylation, which is possible thanks to the 'sophisticated machinery' located in the inner mitochondrial membrane. Mitochondria participate in the regulation of redox state and calcium homeostasis in cell. Cations of calcium regulate some mitochondrial processes, such as enzyme activity, i.e. pyruvate dehydrogenase, or metabolic rate. These organelles participate in biosynthesis of amino acids, vitamin cofactors, fatty acids and neurotransmitters (Waldbaum & Patel, 2009). Many other biochemical reactions are associated with the functioning of these structures, including synthesis of heme group and some steps of steroid synthesis. Also, a part of the processes occurring in the urea cycle take place there (Pinti et al. 2010). Mitochondria have critical function in the control of apoptotic and necrotic cell death and in most types of cells they are also a major site of reactive oxygen species (ROS) generation (Duchen, 2004). ROS are involved in many signaling pathways. Most of them are second messengers that trigger different cellular events, such as cytokine secretion or activation of transcription factors, but in excess they can contribute to the formation of defects in mitochondria, as well as in a whole cell (Edeas et al., 2010a).

3.1 Mitochondrial structure and biogenesis

Mitochondria are encapsulated by two membranes, each with different structure and function, separated by intermembrane space, in which some important proteins involved in the mitochondrial bioenergetics and/or cell death are located (Fig. 3) (Duchen, 2004, Borutaite, 2010). The outer membrane contains porins, which make it permeable to molecules smaller than 5-6 kDa (Waldbaum & Patel, 2009). Compounds such as water, O_2, CO_2, and NH_3 easily pass through the membrane, but hydrophilic metabolites and all inorganic ions in order to get over this membrane require the participation of specific channels and carrier proteins. Such a transport is generally based on the exchange of molecules, i.e. ADP is exchanged for ATP and P_i (inorganic phosphate) for OH⁻ (Szewczyk & Wojtczak, 2002). The mitochondrial inner membrane contains enzymes facilitating an oxidative phosphorylation (OXPHOS). This complex of enzymes consists of four oxidoreductases involved in respiratory electron transport (Complexes I - IV) and the ATP synthase complex (Complex V).

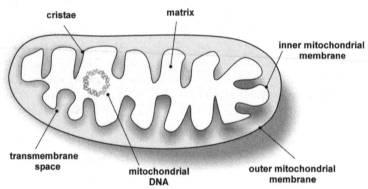

Fig. 3. Mitochondria structure and components.

Until recently, the inner membrane has been described as a multiple infolded structure forming cristae and containing numerous mitochondrial proteins (Duchen, 2004). However, electron tomographic analyses of a variety of mitochondria (both isolated and observed *in situ* in various cell types) have provided overwhelming evidence showing the need of some changes in the perception of the structure of these organelles. These infoldings or rather invaginations are not randomly spaced in the membrane, as often considered, but resemble microcompartments, which face each other in the peripheral region of the membrane. The narrow junctions are wide enough to pass metabolites and many soluble proteins (Mannella, 2008). However, the number of cristae junctions and the morphology of the intercristal space depends on the metabolic state of mitochondria (Logan, 2006). Isolated mitochondria usually occur in one of two morphologic states, condensed or orthodox. The first one is characterized by contracted, very dense matrix and wide cristae. In the second state matrix is expanded and cristae compartments are more compact. Osmotic and metabolic changes in mitochondria are responsible for these alterations. It is believed that mitochondrial inner membrane topology is regulated by the cell to improve mitochondria capacity in their response to stimuli (Mannella, 2008).

Tissue cells contain from a few dozen to several thousands of mitochondria and their number is associated with cell energy demands. Organs such as heart, muscles or brain contain the largest number of mitochondria. Mitochondria are very dynamic structures, which can divide, undergo fusion and can take the form of the network of elongated and interconnected filaments. The phenomena of fission and fusion have an impact on mitochondrial shape, size and number (Logan, 2006). The division and replication of mitochondria is under control of the nucleus and is somehow associated with division and replication of nuclear DNA. Replication of mitochondria requires coordination between the process of mtDNA replication and synthesis of proteins encoded in both genomes (nucleus and mitochondrial). The production of both types of proteins must be synchronized to preserve their functionality.

3.2 Electron transport chain and ATP synthesis

Mitochondrial ATP production involves three main steps: a) the enzymatic "combustion" of acetyl in tricarbolxylic acid cycle (TCA), b) the electron transport chain activity and c) ATP synthase action. Energy released during this cycle is used to reduce the electron carriers NAD^+ to NADH and FAD^{2+} to FADH (Duchen, 2004). Electrons from NADH and $FADH_2$ are transferred to the respiratory chain - a coupled enzyme systems composed of four complexes (Complex I - IV). Complex I (NADH dehydrogenase) is the major entrance point of electrons to respiratory chain and is composed of two domains. One domain, localized in withe membrane, is involved in proton translocation across the bilayer, and the other, matrix-exposed domain, is responsible for oxidation of NADH. $FADH_2$ is the donor of electrons to succinate dehydrogenase (Complex II) which is the second entrance point of electrons to the ETC. Electrons from both complexes are transferred on mobile intermediate – ubiquinone, which is converted to reduced form - ubiquinol. The flow of electrons from ubiquinol is directed through the Complex III, also known as ubiquinol-cytochrome c reductase, to another carrier - cytochrome c, which transfers electrons to Complex IV - cytochrome c oxidase. Finally, at the very end of the respiratory chain, Complex IV reduces the oxygen to water in sequential four-electron transfer (Adam-Vizi & Chinopoulos, 2006). The oxidation of NADH and $FADH_2$ provides the energy to transport protons from mitochondrial matrix into the intermembrane space by the proton pumps (Complexes I, III, IV). The difference in the proton concentration, and thus the difference in the electric charge across the inner mitochondrial membrane creates the electrochemical potential gradient, also called an electrochemical proton gradient or a 'proton-motive force', which is mainly expressed as a mitochondrial transmembrane potential (Nazaret, 2008). The structure of mitochondrial electron transport and the scheme showing ATP production by mitochondria was introduced in Fig. 4.

Energy needed to phosphorylate ADP by ATP synthase comes from the entry of protons back into the matrix through the proton channel of this complex. This process is called oxidative phosphorylation (Frey & Mannella, 2000). ATP is then transported to the cytoplasm by the adenine nucleotide translocase (ANT). However, there are several mechanisms that may lead to the loss of mitochondrial potential, including an inhibition of respiration, failure in substrate supply and uncoupling mechanisms that cause proton leak across the membrane.

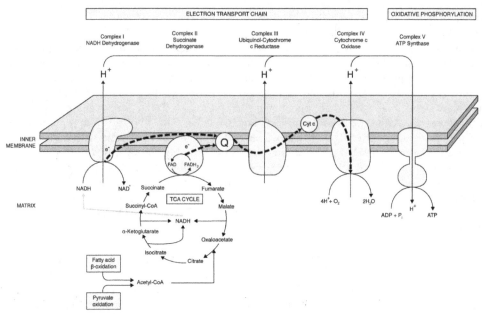

Fig. 4. The general mechanism leading to oxidative phosphorylation is as follows: high-energy electrons (marked as e-) derived from NADH and FADH$_2$, are moving along the respiratory chain composed of four protein complexes (Complex I - IV) and two additional electrons carriers: ubiquinone (coenzyme Q, **Q**), a small molecule freely moving in the inner mitochondrial membrane layer, and cytochrome c (**Cyt c**), localized in the intermembrane space attached to the inner membrane. Part of the energy released in this process is used up in the action of proton pumps transporting protons (H$^+$) from matrix to the intermembrane space. Across the inner membrane electrochemical gradient of protons is formed. Protons tend to return to the mitochondrial matrix and restore alignment of H$^+$ concentration on both sides of the membrane. When they pass back through transmembrane protein complex – ATP synthase - the energy of their movement is used for the synthesis of ATP from ADP and inorganic phosphate (**P$_i$**).

3.3 Free radical generation by mitochondria

Oxidative metabolism and ATP synthesis are closely associated with ROS generation in mitochondria. These organelles consume 80-90% of cell's oxygen during oxidative phosphorylation. The electron transport chain is the main source of ROS in functioning mitochondria. Approximately 0.2-2% of the oxygen taken up by a cell is converted by mitochondria to ROS. Superoxide (O2$^{\bullet-}$) is the main product of these transformations, and it is then converted to hydrogen peroxide (H$_2$O$_2$) by spontaneous dismutation or by superoxide dismutase (SOD). Glutathione peroxidase or catalase, in turn, convert hydrogen peroxide into water. If this change does not occur, in the presence of divalent cations H$_2$O$_2$ can undergo Fenton's reaction to produce even more harmful hydroxyl radical ($\bullet$OH). Oxygen can be reduced to superoxide in one-electron step, theoretically, at each step of the respiratory chain, but in reality two major sites of superoxide generation are Complex I and Complex III (Paradies et al., 2010, as cited in Murphy, 2009). There is a considerable experimental support for two mechanisms of ROS production by complex I. The first one is the production of ROS as a consequence of so-called reverse electron transfer (RET) in the mitochondrial respiratory chain. RET is a set of redox reactions in the mitochondrial ETC that allows electrons to flow from coenzyme Q to NAD^+ instead to oxygen. The other one takes place under normal conditions, whereas most of the energy from the creation of mitochondrial potential difference is used to generate ATP through ATP synthase. This process causes collapse of the proton gradient. The amplitude of the electrochemical proton gradient regulates the flow of electrons through the ETC. When the electrochemical potential gradient is high, for instance under conditions of high glucose concentrations, the life of electron transport intermediates that are involved in superoxide formation, such as ubisemiquinone, is prolonged. The reason of such condition is that the activities of of ETC proton pumps depend on the proton gradient across the inner membrane and the membrane itself – two components of proton-motive force (Duchen, 2004).

3.4 Free radical targets and the oxidative vicious circle

Mitochondria are continuously exposed to action of reactive oxygen species so they need to have a system that will prevent them against destructive effect of oxidative damage. In fact, mitochondria are equipped in complicated multi-leveled ROS defense network consisting of enzymes and non-enzymatic antioxidants. They contain a high concentration of glutathione, α-tocopherol and manganese-containing superoxide dismutase (MnSOD). The role of MnSOD is the dismutation process of superoxide radical to H_2O_2. The product of MnSOD reaction is detoxified by other enzymes, i.e. catalase, which converts H_2O_2 into O_2 and H_2O. Mitochondria possess also another system capable of efficient superoxide removal - the cytochrome c, which is than regenerated (oxidized) by its natural electron acceptor, cytochrome c oxidase (Complex IV). In intact mitochondria, superoxide may be efficiently scavenged by intramitochondrial antioxidant defences (Duchen, 2004). An imbalance between oxidants and antioxidants induces oxidative stress responsible for alteration of biomolecules and intracellular signaling pathways present in every cell (Edeas et al., 2010a). Mitochondria are a major source of ROS generation, but what is important, they are also its major target (Duchen, 2004). Mitochondrial membrane lipids, mainly long-chain polyunsaturated fatty acids (PUFAs), are also susceptible to oxidative stress. PUFAs are basal components of mitochondrial phospholipids. The sensitivity of PUFAs to oxidation increase with the increasing number of double bonds per fatty acid molecule. Peroxidation of membrane phospholipids causes alterations in their structure and consequently may disrupt organisation of the lipid bilayer. It contributes also to changes in membrane fluidity and/or permeability, and causes changes in the mitochondrial membrane potential, in respiratory capacity and in oxidative phosphorylation. ROS are also responsible for alterations in proteins, which may manifest by changes in their structure, proteolytic susceptibility and spontaneous fragmentation. Oxidative damage especially affects the mitochondrial electron transport chain and, when the ETC enzymes stop working properly, the ROS production increases. This may result in the incomplete oxygen consumption, reduced production of ATP, and finally overproduction of ROS (Waldbaum & Patel, 2009).

4. Mitochondrial physiology in diabetes

Mitochondria are provided with a variety of bioenergetic functions mandatory for the regulation of intracellular energy production. Alteration of bioenergetic activities may have drastic consequences on cellular function through the perturbation of energetic charge and balance of the cell (Fig. 5).

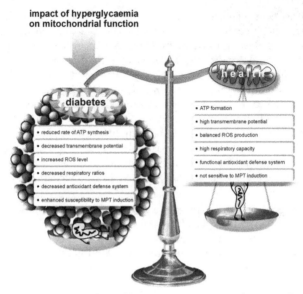

Fig. 5. The overall impact of the burden of hypergycaemia in diabetes on functioning of mitochondria.

Abnormalities of mitochondrial metabolism causing human disease have been recognised for more than 40 years. Numerous reports clearly indicate the association between mitochondrial dysfunction and diabetes. Nevertheless, some mechanisms of mitochondrial role in this pathology still requires further elucidation. Therefore different animal model studies are involved in the investigation of explaining these unknown mechanisms. There are several models of experimental diabetes that mimic two common types of diabetes. Streptozotocin-induced diabetes is a widely accepted animal model for type 1 diabetes, resulting from the inability of the pancreatic beta cells to produce insulin. For the research on type 2 or the insulin resistant state, resulting from the inefficient use of insulin by the tissues to regulate blood glucose concentration, some genetically manipulated animal models (e.g. Zucker fatty rats (ZFR), *ob/ob* (obese) mice, CP (corpulent) rats, GK (Goto-Kakizaki) rats, Akita mice) may be utilized (Srinivasan & Ramarao, 2007).

4.1 Mitochondrial dysfunction and diabetes type 1

Streptozotocin (STZ) is a naturally occurring chemical that is particularly toxic to the insulin-producing β cells of the pancreas in mammals and is used to generate Type 1 diabetes in the experimental model. Animals with diabetes induced by STZ exhibit increased mitochondrial oxidative stress and dysfunction. Other agent, alloxan, a toxic glucose analogue, is also used in order to generate type 1 of diabetes. Alloxan selectively destroys insulin-producing cells in the pancreas when administered to rodents and many other animal species, and has been shown to cause also mitochondrial dysfunction. Alloxan-treated severe diabetic rats were shown to exhibit impaired mitochondrial phosphorylative activities and low mitochondrial oxidation-reduction states (Yamamoto et al., 1981). In one month old alloxan-diabetic animals the enzyme activity of the mitochondrial membrane marker, F_0F_1-ATPase, was found to be decreased. Insulin treatment caused hyperstimulation of the activity, whereas in late-stage diabetes the catalytic efficiency of the enzyme was increased and became decreased upon insulin treatment (Patel & Katyare, 2006).

Mitochondrial dysfunction in diabetic rats can be succinctly summarized into: decreased mitochondrial 3'-AMP forming enzyme activity, increased oxidative and nitrosative stress, decreased oxygen consumption, loss in mitochondrial transcriptional capacity, increased HMG-CoA synthase, increased levels of pyruvate and dicarboxylate transporters, increased degradation of ATPase, changes in phospholipid composition, increased pyruvate carboxylase activity, increased fatty acid beta oxidation and ultrastructure alterations.

4.2 Mitochondrial dysfunction and diabetes type 2

Type 2 diabetes is the most common metabolic disease in the world, and its prevalence much exceeds the prevalence of type 1 diabetes. Among different causes leading to diabetes the role of mitochondria is considered substantial. Disorders of the mitochondrial electron transport chain, overproduction of ROS and lipoperoxides or impairments in antioxidant defenses are encountered in type 2 diabetes. Increased ROS levels lead to generalized oxidative damage to all mitochondrial components. Moreover, it is well established that mitochondrial function is required for normal glucose-stimulated insulin secretion from pancreatic β cells. However, the studies in humans suggest that more subtle defects in mitochondrial function may also play a role in the pathogenesis of insulin resistance and type 2 diabetes (Fig. 6) (Luft, 1994).

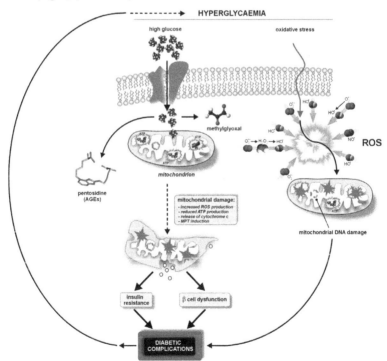

Fig. 6. Relationships among hyperglycaemia, mitochondrial damage, oxidative burst and diabetic complications.

Some data support the hypothesis that insulin resistance in humans arises from defects in mitochondrial fatty acid oxidation, which in turn leads to increased intracellular fatty acid metabolites that disrupt insulin signaling (Petersen et al., 2003). Alternatively, the reduction in mitochondrial oxidative phosphorylation activity in insulin-resistant individuals could be due not to mitochondrial loss, but rather to a defect in mitochondrial function. This hypothesis is supported by muscle biopsy studies. In one such study, the activity of mitochondrial oxidative enzymes was found to be lower in type 2 diabetic subjects, and in another, the activity of mitochondrial rotenone-sensitive nicotinamide adenine dinucleotide oxidoreductase [NADH:O(2)] was found to be lower (Lowell & Shulman, 2005).

4.3 Alterations in cardiac mitochondria observed in diabetes

Cardiovascular diseases are the predominant cause of death in patients with diabetes mellitus. Underlying mechanism for the susceptibility of diabetic patients to cardiovascular diseases still remains unclear. Elevated oxidative stress was detected in diabetic patients and in animal models of diabetes. Hyperglycaemia, oxidatively modified atherogenic lipoproteins, and advanced glycation end products act in a concerted action together with oxidative stress, and cumulatively contribute to progression of late diabetic complications. Mitochondrial dysfunction increases electron leak and the generation of ROS from the mitochondrial respiratory chain (MRC). High levels of glucose and lipids impair the activities of MRC complex enzymes. Furthermore, increased activity of NADPH oxidase (NOX), which generates superoxide from NADPH in cells, was detected in diabetic patients (Shen, 2010).

Because mitochondria constitute 20–30% of the cardiac myocytes, one of the potent causes for heart malfunctioning in diabetes is the impaired mitochondrial function and consequently the decreased ATP generation (Rolo & Palmeira 2006).

Many reports evidenced that diabetic hearts show impaired mitochondrial function, decreased ATP generation, decreased oxidative capacity, increased ROS, abnormal morphology, increased UCP-3 level, decreased mitochondrial calcium uptake and increased susceptibility to MPT induction (Fig. 5).

Distortions in cardiac mitochondrial bioenergetics are known to occur in both human types of diabetes and in models of diabetes in animals. Reduced mitochondrial calcium uptake was observed in heart mitochondria from STZ-treated rats. This was related to enhanced susceptibility to MPT induction rather than damage to the calcium uptake machinery. Interestingly, heart mitochondria from GK rats were less susceptible to the induction of MPT, showing larger calcium accumulation before the overall loss of mitochondrial impermeability. Different approaches of antioxidant administration in GK rats (vitamin E or coenzyme Q_{10}) showed no success in reversing the diabetic phenotype (Oliveira et al., 2003).

Diabetic heart failure may be causally associated with alterations in cardiac energy metabolism. Fuel selection and capacity for ATP production in the normal and failing heart are dictated by several metabolic regulatory events at the level of gene expression. Decline in the capacity for ATP, as caused by progressive impairment of mitochondrial function, is a gradual step in the progression to heart failure of any cause. Fetal heart depends on glucose and the adult heart on glucose and fatty acids. The switch between fatty acid oxidation and glucose in the adult heart leads to a healthy metabolic situation (Huss & Kelly, 2005). In the insulin-resistant and diabetic heart, fatty acid oxidation is increased and glucose utilisation is diminished. Long-term consequence of fatty acid oxidation is mitochondrial dysfunction. A number of mechanisms may be responsible for enhanced fatty acid utilisation in type 2 diabetic hearts, such as increased fatty acid uptake into the cell and mitochondria, increased UCP-3 expression, and stimulation of peroxisome proliferator-activated receptor-α (PPARα) (Fig. 7) (Rolo & Palmeira, 2006).

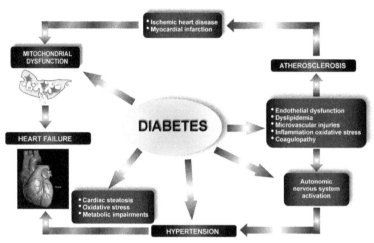

Fig. 7. The role of cardiac mitochondria in the development of heart failure in a course of diabetes.

Diabetes-associated metabolic disorders may cause the mitochondrial dysfuntion and upregulation of NOX in the cardiovascular system, which lead to increased ROS production and oxidative stress in vasculature and blood circulation. ROS may directly oxidize or indirectly regulate molecules related to atherosclerosis and thrombosis. Mitochondrial NOX, or its regulators may be considered as potential drug targets for the prevention and/or treatment of diabetic cardiovascular complications.

5. Therapeutic approaches to reduce diabetic complications

Patients with diabetes mellitus are usually treated with a combination of pharmacological agents and habitual approach, i.e. their lifestyle modification. The development of new antidiabetic agents, such as insulin analogs and incretin-based therapies, has led to treatment strategies that enable numerous patients with diabetes to improve their lifestyles.

5.1 Use of insulin in a common diabetes therapy – its advantages and disadvantages

Type 1 diabetes mellitus is underlied by the shortage of insulin, which plays a crucial role of carbohydrate and fat metabolism. Its absence causes rather a complex array of serious impairments in patients' health, e.g. hiperglycaemia, ketoacidosis, coma and even death. Hence, the injection of exogenous insulin to diabetic individual is essential to: (a) maintain a normal glucose concentration and (b) avoid the advanced microvascular complications, such as retinopathy, nephropathy or neuropathy.

Insulin was discovered by Banting and Best and this event was a milestone in the treatment of patients with diabetes (Bliss, 1982). Initially, bovine, porcin and even some fish analogues were applied to avoid diabetic ketoacidosis. However, gradually the scientists were challenged by the uprising problems in patients injected with animal insulins (mainly because of their rather high impurity and immunogenicity) and were forced to develop a new class of insulins. Improved techniques used for insulin purification combined with other compounds like protamine and zinc, enabled to manufacture protamine insulin with the prolonged time of activity and later, the more stable protamine zinc insulin (Hagedorn et al., 1936). Moreover, further scientific discoveries shed light on better understanding of insulin structure and activity and initiated a new avenue to design human insulin analogues characterized by the properties of prolonged hormone activity in a bloodstream.

At present, there are few types of short-acting insulin analogues used in anti-diabetes therapy, and among them:

- **insulin lispro** (Humalog manufactured by Elli Lilly and Company), which was approved and launched into the market in 1996. Its modified amino acid sequence provides faster absorption, which is essential to ameliorate postprandial glucose level. The studies revealed that the activity peak of lispro appears in 1 hour after using and lasts for next 3-4 hours (Howey et al., 1994)
- **insulin glulisine** (Apidra manufactured by Sanofi-Aventis), which possesses asparagine at position B3 and glycine at position B29 in amino acid chain (Garg et al., 2005).
- **insulin aspart** (NovoRapid manufactured by Novo Nordisk), in which proline is replaced with aspartic acid what facilitates its faster absorption (Mudaliar et al., 1999).

Long-acting insulin analogues are crucial to mimic the endogenous insulin secretion. Thereby a specific modification of insulin structure was essential to obtain longer acting analogues. There are two approaches leading to diminish absorption: the first one is to change the isoelectric point of insulin and the second one is to acetylate a hydrophobic residue with fatty acid.

- **insulin glargine** (Lantus, Sanofi-Aventis) exemplifies a long-acting insulin analogue, in which asparagine is substituted by glycine at position A21 and the position B30 is enriched with two molecules of arginine at B31 and B32 (Bolli & Owens, 2000). These alterations have an impact on the structure and an isoelectric point of insulin, contributing to a decrease in its solubility after injection. Finally, the result of these changes is the product, which acts about 20 hours (Heise et al., 2002).
- **insulin detemir** (Levemir, Novo Nordisk) is characterized by long acting properties (17-20 hours) obtained as a result of acetylation with fatty acid at the position B29 and by removal of threonine at B30 (Havelund et al., 2004).

After long years of experience and observations, nowadays, multiple daily injection program is believed to be the most reasonable approach in modern diabetic treatment in order to mimic the physiological insulin release. However, it implies that patients have to undertake more inconvenient therapy resulting from the scheduled injections of both short- and long-acting insulins. To deal with this problem, clinicians and patients may choose an alternative method, which requires biphasic insulin analogues administration.

Patients with pre-diagnosed type 2 diabetes should take seriously into account the radical change of their lifestyle in order to cause a delay of possible medical intervention. Under conditions when diet or changing a lifestyle may not be sufficient enough, additional pharmacological treatment is required to avoid a severe consequence of this disease.

Nowadays, medicine is focused on delivery a large number of drugs. From pharmacological point of view, these compounds should be effective in improving insulin efficiency or effective in enhancing its secretion from pancreas. Among those substances are: sulfonylureas, biguanides, thiazolidinediones, meglitinides, α-glucosidase inhibitors, amylin analogues, incretin hormone mimetics and dipeptidyl peptidase 4 inhibitors.

- **sulfonylureas** belong to the drugs most frequently used in diabetes treatment. It is known that these oral hypoglycaemic agents interact with β-cell pancreas cells causing insulin secretion, insulin sensitivity amelioration, as well as glucose synthesis reduction. However, in order to treat patients using sulfonylureas, an endogenous secretion of insulin at the same/similar level and a balanced diet are required (Gerich, 1989)
- **metformin**, representing a class of biguanides, is a commonly used oral hypoglycemic agent for the treatment of type 2 diabetes. Metformin is also known as an inhibitor of high glucose- or AGEs-induced ROS generation (Bellin et al., 2006)
- **thiazolidinediones (glitazones, TZD)**, oral anti-diabetic drugs, are based on the improving of adipose tissue and muscle sensitivity to insulin treatment (Day, 1999). However, troglitazone, one of the class of thiazolidinediones, was taken off the market since the hepatotocity has been noticed (Watkins & Whitcomb, 1998).

Development of the obesity associated with diabetes requires using a novel combination treatment, which aims at retarding the microvascular and macrovascular complications occurring in diabetes and obesity. Therefore, some anti-diabetic agents have been indicated to maintain an adequate glucose level in an organism suffering from diabetes. The number of anti-diabetic drugs delivered by subcutaneous injection increases constantly and to date there are numerous agents already launched into market, as well as others, tested in the clinical trials (Fig. 8).

- **glucagon-like-peptide-1 agonist (GLP-1)**, one of the first among subcutaneous drugs used in medicine. Its beneficial effect was achieved by suppression of glucagon secretion and weight loss. Unfortunately, it soon appeared that in the organism GLP-1 was active only 2 min. It was the main reason why scientists started to work on improved GLP-1 analogues.
- **exenatide and liraglutide** were introduced into market in the year 2005 and 2009, respectively, as novel GLP-1 analogues. Each of them decreases HbA_{1c} level and provides the weight loss. The main difference between these agents is their half-lives in a circulation. Exenatide yields therapeutic effect in 4-6 hours, whereas half-life of liraglutide was enhanced to 12-15 hours (Gentilella et al., 2009).
- **bromocriptin** belongs to drugs used in the treatment of Parkinson disease. However, recently it has been shown that this therapeutic agent ameliorates insulin sensitivity, leading to enhancing glucose control and lowering the incidence of hypoglycaemia (Pijl et al., 2000). Although bromocriptin provides weight loss and diminishes concentration of plasma triglyceride and free fatty acids, it was evidenced that after using bromocriptin the patients suffer from several side effects, like nausea, hypotension and psychiatric disturbances.

Diabetes mellitus has been associated with the increased mortality risk due to non-diabetic factors, like several types of solid tumours, including the cancers of colon, breast and pancreas. Similar associations have been noted for central obesity and other conditions associated with increased levels of circulating insulin. These observations have given rise to the hypothesis that growth of these tumours, which are characterised by abnormal expression and function of the insulin–IGF-1 series of receptors, may be promoted by the trophic action of insulin interacting with these receptors. The cancer risk associated with diabetes may also be influenced by therapy in a given diabetic individual: for example, the risk of colon cancer is higher in individuals on insulin, patients on metformin are less likely to be diagnosed with cancer, and the risk of mortality from solid tumours is lower for metformin than for exogenous insulin or sulfonylureas. As a recognition dawns that cancer should be numbered among the complications of diabetes, the possibility that therapies for diabetes may influence tumour progression is likely to attract the increasing interest and concern. Furthermore, the observation that both endogenous insulin and exogenous insulin therapy are associated with tumour progression raises the questions as to the safety of the insulin analogues, which have subtly modified receptor binding properties and may accelerate the growth and proliferation of both healthy and tumour cell lines in culture.

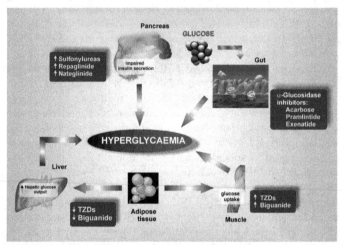

Fig. 8. Sites of action of major oral therapeutical agents used in the treatment of type 2 diabetes. Pharmacological therapies aimed at: inhibiting carbohydrate breakdown in the gut (α-glucosidase inhibitors), stimulating insulin secretion (sulfonylureas, repaglinide, nateglinide), suppressing hepatic gluconeogenesis (thiazolidinediones, biguanide), or accelerating skeletal muscle glucose metabolism (thiazolidinediones, biguanide) exhibit beneficial effects on fasting and/or postprandial plasma glucose, and consequently concord overall metabolic control in type 2 diabetic patients. Thus, a need for concerted combination therapy to successfully control a burden of hyperglycaemia in a majority of DM2 patients becomes a growing expectation by physicians and other health caregivers.

5.2 New agents – new hopes and new perspectives

The growing number of people with diabetes still requires novel combined treatments aimed at retardation of microvascular and macrovascular complications in the future. Therefore, anti-diabetic complications agents are sought to maintain an adequate glucose level in an organism. The number of anti-hyperglycaemic drugs delivered by subcutaneous injection increases constantly and to date there are numerous agents launched into market and examined in clinical trials. However, no such compounds/drugs that could be successfully applied in the treatment of diabetes have emerged hitherto, as the validated outcomes of clinical trials. Intensive studies are continued in order to develop a modern formula for effective amelioration of a burden associated with diabetes and late diabetic complications in diabetic patients in the future.

Studies on the formation of AGEs have been conducted with the goal to find promising pharmacological agents used in prevention or curing diabetic complications. The main target for these agents is to retard the formation of AGEs or to brake the AGE cross-links formed during Maillard reaction.

In order to prevent the AGEs formation the following therapeutic inhibitors have been developed and studied:

- **pyridoxamine (PM)**, a form of vitamin B_6, is thought to inhibit AGEs structures by capturing redox metal ions (Voziyan et al., 2003). In order to examine the anti-diabetic properties of PM under *in vivo* conditions, the model of diabetes mellitus induced by streptozotocin (STZ) was applied. The results of PM administration demonstrated the reduction in hyperglycaemia level and improvement in the plasma lactate/puryvate ratio. The decreased amounts of AGEs have also been noted (Degenhard et al., 2002). Moreover, it was also revealed that PM can act as inhibitor of protenuria and hyperlipidaemia in diabetes mellitus type 1 (Voziyan, 2005).

- **benfotiamine** is a derivative of vitamin B_1, which is involved in a limitation of methylglyoxal formation and lowering of AGEs accumulation (Gadau et al., 2006).

- **ALT-711** is a new stable analogue of N-phenal thiazolinium bromide (PTB). Its mechanism of action is based on preventing metal-catalyzed glycation (Price et al., 2001).

- **Acetylsalicylic acid (Aspirin®), salicylates and ibuprofen** possess an anti-inflammatory properties, which are important in the decreasing of the risk of cataract in people suffering from diabetes. As a radical scavengers, they may reduce the level of free radicals and/or chelate metal ions (Dinis et al., 1994).

- **chromium** deficiency is associated with the blood sugar irregularities of diabetes. Recent studies have demonstrated that chromium is effective in treating various types of diabetes, including types 1 and 2, gestational, and steroid-induced diabetes. Treatment of type 2 diabetes with chromium has led to improvement in blood glucose, insulin, and haemoglobin A_{1c} (HbA_{1c}) levels. The use of organic chromium complexes has been found to give superior results when compared to inorganic salts. Chromium has been found to be effective in reversing diabetes caused by the therapeutic use of glucocorticoids. Chromium picolinate (600 μg/day) was effective in lowering blood glucose almost twice (from 13.9 mM/L to 8.3 mM/L) in 47 of 50 patients. This therapy in patients was also able to reduce the doses of insulin and/or hypoglycaemic medications by half within one week from the beginning of chromium supplementation (Lamson & Plaza, 2002)

Among the newest agents tested in both *in vitro* and *in vivo* studies are poly(amido)amine PAMAM dendrimers and β-resorcylidene aminoguanidine (RAG), a derivative of aminoguanodine.

- **PAMAM dendrimers** are widely studied all over the world in almost every field of science. Their unique structure with nucleophilic character provided by surface amino groups may play an important role in the prevention or amelioration of hyperglycaemia. The ability of conjugation of PAMAM dendrimers to certain biologically relevant molecules makes them promising agents for using in biomedicine area, either as drugs or drug delivery systems. Experimental *in vitro* studies revealed that dendrimers appear very effective in scavenging glucose and reducing protein glycation (Fig. 9) (Labieniec & Watala, 2010). It was also evidenced that PAMAM dendrimer G4 administrated to rats with streptozotocin-diabetes acted as glucose scavenger and suppressed the accumulation of AGEs products, as well as some other markers of oxidative and carbonyl stress (Labieniec et al., 2008).

- **aminoguanidine** was the most promising oral antihyperglicaemic agent based on antioxidant capability and reduction of carbonyl reactive intermediates (Brownlee et al., 1986). Nevertheless, the B_6 vitamin depletion and oxidative stress production after using of aminoguanidine by patients with diabetes has been recorded, and therefore this compound was removed from clinical trials. Since then, new analogues of aminoguadine were synthesized in order to avoid the undesirable side effects of aminoguanidine itself. **β-resorcylidene aminoguanidine (RAG)** is one of these analogues, which seems to be the most promising and effective antioxidative and anti-diabetic agent amongst the others tested hitherto. The majority of studies have demonstrated that RAG is able to limit diabetes-associated long-term complications (protein glycation, AGEs formation, ROS level). Scientists suggest that RAG acts not only as antioxidative and/or anti-diabetic agent (Waczulíková et al., 2000, Vojtaśśak et al., 2008), but has also been shown to act as antithrombotic compound, independently of its anti-glycation activities (Watala et al., 2009).

A   **B**

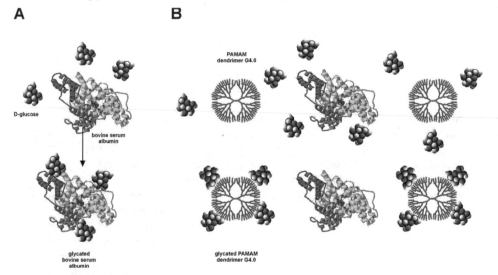

Fig. 9. The proposed mechanism of anti-glycation action of poly(amido)amine dndrimers (PAMAM).

Under conditions of excessive glucose the model protein (bovine serum albumin, BSA) undergoes extensive glycation (A), which becomes retarded and reduced to a large extent in the presence of poly(amido)amine dndrimers, generation 4.0 (PAMAM G4.0).

6. Conclusion

Diabetes is not merely a disease of impaired insulin sensitivity or insulin release, but may be a global metabolic dysfunction, including, among others, the collapse of the mitochondrial energy system. The role of the mitochondria in the metabolism associated with the pathophysiology of diabetes seems unique, mainly because a generation of ROS (which seem a natural part of mitochondrial physiology) constitutes a major treat in the development of diabetic sequelae. ROS play the central role in mediating various metabolic defects associated with the diabetic state. Therefore, inhibition of ROS production and/or enhancement of ROS scavenging will prove to be beneficial therapies. Hyperglycaemia elicits an increased ROS production, presumably from the mitochondrial respiratory chain. An important challenge for future research is to determine whether strategies aimed to improving mitochondrial functionality by using agents with anti-diabetic properties might have therapeutic potential in the treatment of diabetes. On the other hand, the better understanding of mitochondrial biology is still needed to facilitate the judicious selection and development of compounds/agents, which could be used as "mitochondrial drugs". Further studies are certainly required to better understand how these novel compounds and mitochondria may interact with each other, and how our understanding of such interaction might be utilized for the impaired mitochondrial functioning in the presence of diabetes. These investigations should also determine, which genetic, environmental, pharmacological and nutritional factors are possibly involved in an individual patient's susceptibility and, which treatments can be used safely in those patients, who suffer from heavy diabetes and are crushed by the burden of advanced long-term complications.

7. References

Adam-Vizi, V. & Chinopoulos, C. (2006). Bioenergetics and the formation of mitochondrial reactive oxygen species. *Trends Pharmacol Sci*, Vol. 27, No. 12, n.d., pp. 639-645

Barnaby, O.S.; Cerny, R.L., Clarke, W. & Hage, D.S.(2011). Comparison of modification sites formed on human serum albumin at various stages of glycation. *Clin Chim Acta*, Vol. 412, No. 3-4, (January 2011), pp. 277-285

Bellin, C.; de Wiza, D.H.; Wiernsperger, N.F. & Rosen, P. (2006). Generation of reactive oxygen species by endothelial and smooth muscle cells: influence of hyperglycemia and metformin. *Horm Metab Res*, Vol. 38, No. 11, (November 2006), pp. 732-739

Bliss, M. (1982). Banting's, Best's, and Collip's accounts of the discovery of insulin. *Bull Hist Med*, Vol. 56, No. 4, n.d., pp. 554-568

Bolli, G.B. & Owens, D.R. (2000). Insulin glargine. *Lancet*, Vol. 356, No. 9228, (August 2008), pp. 443-445

Bonnefont-Rousselot, D. (2002). Glucose and reactive oxygen species. *Curr Opin Clin Nutr Metab Care*, Vol. 5, No. 5, (September 2002), pp. 561-568

Borutaite V. (2010). Mitochondria as decision-makers in cell death. *Environ Mol Mutagen*, Vol. 51, No. 5, (March 2010), pp. 406-416

Brownlee, M.; Vlassara, H.; Kooney, A.; Ulrich, P. & Cerami, A. (1986). Aminoguanidine prevents diabetes-induced arterial wall protein cross-linking. *Science*, Vol. 232, No. 4758, (June 1986), pp. 1629-1632

Chiu, C.J. & Taylor. A. (2011). Dietary hyperglycemia, glycemic index and metabolic retinal diseases. *Prog Retin Eye Res.* Vol. 30, No. 1, (January, 2011), pp.18-53

Dinis, T.C.; Maderia, V.M. & Almeida, L.M. (1994). Action of phenolic derivatives (acetaminophen, salicylate, and 5-aminosalicylate) as inhibitors of membrane lipid peroxidation and as peroxyl radical scavengers. *Arch Biochem Biophys*, Vol. 315, No. 1, (November 1994), pp.161-169

Duchen, M.R. (2004). Mitochondria in health and disease: perspectives on a new mitochondrial biology. *Mol Aspects Med*, Vol. 25, No. 4, (August 2004), pp. 365-451

Edeas, M.; Attaf, D.; Mailfert, A.S; Nasu, M. & Joubet, R. (2010a) Maillard Reaction, mitochondria and oxidative stress: Potential role of antioxidants. *Pathologie Biologie*, Vol. 58, No. 3,(June 2010), pp. 220–225

Edeas, M. & Robert. R. (2010b). The Maillard reaction, its nutritional and physiopathological aspects. Introduction. *Pathol Biol, Vol.* 58, No. 3, (June 2010), pp.199

Ellis, G.P. (1959). The Maillard reaction. *Adv Carbohydr Chem, Vol.* 14, n.d., pp. 63-134

Frey, T.G. & Mannella C.A. (2000). The internal structure of mitochondria. Trends *Biochem Sci*, Vol. 25, No. 7, (July 2000), pp. 319-324

Gadau, S.; Emanueli, C.; Van Linthout, S.; Graiani, G.; Todaro, M.; Meloni, M.; Campesi, I.; Invernici, G.; Spillmann, F.; Ward, K. & Madeddu, P. (2006). Benfotiamine accelerates the healing of ischaemic diabetic limbs in mice through protein kinase B/Akt-mediated potentiation of angiogenesis and inhibition of apoptosis. *Diabetologia*, Vol. 49, No. 2, (February 2006), pp. 405-420

Garg, S.K.; Ellis, S.L. & Ulrich, H. (2005). Insulin glulisine: a new rapid-acting insulin analogue for the treatment of diabetes. *Expert Opin Pharmacother*. Vol. 6, No.4,(April 2005), pp. 643-651

Gerich, J.E. (1989). Oral hypoglycemic agents. *N Engl J Med*, Vol. 321, No. 18, (November 1989), pp.1231-1245

Goldberg, T.; Cai, W.; Peppa, M.; Dardaine, V.; Baliga, B.S.; Uribarri, J. & Vlassara. H. (2004). Advanced glycoxidation end products in commonly consumed foods. *J Am Diet Assoc*, Vol. 104, No. 8, (August 2004), pp.1287-1291

Heise, T.;. Bott, S.; Rave, K.; Dressler, A.; Rosskamp, R. & Heinemann, L. (2002). No evidence for accumulation of insulin glargine (LANTUS): a multiple injection study in patients with Type 1 diabetes. *Diabet Med*, Vol. 19, No. 6, (June 2002), pp. 490-495

Hagedorn, H.C. (1937). Protamine Insulinate: (Section of Therapeutics and Pharmacology). *Proc R Soc Med*, Vol. 30, No. 6, (April 1937), pp. 805-814

Harman, D. (1956). Aging: a theory based on free radical and radiation chemistry. *J Gerontol*, Vol. 11, No. 3, (July, 1956), pp. 298-300

Havelund, S.; Plum, A.; Ribel, U.; Jonassen, I.; Volund, A.; Markussen, J. & Kurtzhals, P. (2004). The mechanism of protraction of insulin detemir, a long-acting, acylated analog of human insulin. *Pharm Res*, Vol. 21, No. 8, (August 2004), pp. 1498-1504

Hodge, J. E. (1955). The Amadori rearrangement. *Adv Carbohydr Chem*, Vol.10, n.d., pp.169-205

Howey, D.C.; Bowsher, R.R;. Brunelle, R.L. & Woodworth, J.R. (1994). [Lys(B28), Pro(B29)]-human insulin. A rapidly absorbed analogue of human insulin. *Diabetes*, Vol. 43, No. 3, (March 1994), pp. 396-402

Hunt, J.V.; Dean, R.T. & Wolff, S.P. (1988). Hydroxyl radical production and autoxidative glycosylation. Glucose autoxidation as the cause of protein damage in the experimental glycation model of diabetes mellitus and ageing. *Biochem J*, Vol. 256, No. 1, (November, 1988), pp. 205-212

Huss, J. M. & Kelly, D.P. (2005). Mitochondrial energy metabolism in heart failure: a question of balance. *J Clin Invest*, Vol. 115, No. 3, (March, 2005), pp. 547-555

Jaeger, H.; Janositz, A. & Knorr, D. (2010). The Maillard reaction and its control during food processing. The potential of emerging technologies. *Pathol Biol*. Vol. 58, No. 3, (June 2010), pp. 207-213

King, H. & Rewers, M. (1993). Global estimates for prevalence of diabetes mellitus and impaired glucose tolerance in adults. WHO Ad Hoc Diabetes Reporting Group. *Diabetes Care*, Vol. 16, No. 1, (January, 1993), pp.157-177

King, H.; Aubert, R.E. & Herman, W.H. (1998). Global burden of diabetes, 1995-2025:prevalence, numerical estimates, and projections. *Diabetes Care*, Vol. 21, No. 9, (September 1998), pp. 1414-1431

Kuznetsov, A. V. & Margreiter, R. (2009). Heterogeneity of mitochondria and mitochondrial function within cells as another level of mitochondrial complexity. *Int J Mol Sci*, Vol. 10, No. 4, April 2009, pp. 1911-1929, ISSN 1422-0067

Labieniec, M.; Ulicna, O.; Vancova, O.; Glowacki, R.; Sebekova, K.; Bald, E.; Gabryelak, T. & Watala, C. (2008). PAMAM G4 dendrimers lower high glucose but do not improve reduced survival in diabetic rats. *Int J Pharm*, Vol. 364, No. 1, (November 2008), pp. 142-149.

Labieniec, M. & Watala C. (2010). Use of poly(amido)amine dendrimers in prevention of early non-enzymatic modifications of biomacromolecules. *Biochimie*, Vol. 92, No. 10, (October 2010), pp. 1296-1305

Lamson, D.W. & Plaza, S.M. (2002). The safety and efficacy of high-dose chromium. *Altern Med Rev*, Vol. 7, No. 3, (June 2002), pp. 218-235

Logan, D.C. (2006). The mitochondrial compartment. *J Exp Bot*, Vol. 5, No. 6, March 2006, pp. 1225-1243

Lowell, B.B. & Shulman, G.I. (2005). Mitochondrial dysfunction and type 2 diabetes. *Science*, Vol. 307, No. 5708, (January 2005), pp. 384-387

Luft, R. (1994). The development of mitochondrial medicine. *Proc Natl Acad Sci U.S.A*, Vol. 91, No. 19, (September 1994), pp. 8731-8738

Mannella, C.A. (2008). Structural Diversity of Mitochondria: Functional Implications. *Ann NY Acad Sci*, Vol. 1147, pp. 171–179

Marko, D.; Habermeyer, M.; Kemeny, M.; Weyand, U.; Niederberger, E., Frank, O. & Hofmann, T. (2003). Maillard reaction products modulating the growth of human tumor cells in vitro. *Chem Res Toxicol.* Vol. 16, No. 1, (January, 2003), pp. 48-55

Miyata, T.; Sugiyama, S.; Saito, A. & Kurokawa, K. (2001). Reactive carbonyl compounds related uremic toxicity ("carbonyl stress"). *Kidney Int Suppl,*Vol. 78, (February 2001), pp. S25-S31

Mudaliar, S R.; Lindberg, F.A.; Joyce, Beerdsen, M.P.; Strange, P.; Lin, A. & Henry, R.R. (1999). Insulin aspart (B28 asp-insulin): a fast-acting analog of human insulin: absorption kinetics and action profile compared with regular human insulin in healthy nondiabetic subjects. *Diabetes Care*, Vol. 22, No. 9, (September 1999), pp. 1501-1506

Nazaret, C.; Heiske, M.; Thurley, K. & Mazat, J.P. (2008). Mitochondrial energetic metabolism: A simplified model of TCA cycle with ATP production. *J Theor Biol*, Vol. 258, No. 3, (June 2009), pp. 455-464

Oliveira, P.J.; Rolo, A.P.; Seica, R.; Santos, M.S.; Palmeira, C. M. & Moreno, A. J. (2003). Reduction in cardiac mitochondrial calcium loading capacity is observable during alpha-naphthylisothiocyanate-induced acute cholestasis: a clue for hepatic-derived cardiomyopathies? *Biochim Biophys Acta*, Vol. 1637, No. 1, (January 2003), pp. 39-45

Paradies, G.; Petrosillo, G.; Paradies, V. & Ruggiero, F.M. (2010). Oxidative stress, mitochondrial bioenergetics, and cardiolipin in aging. *Free Radic Biol Med*, Vol. 48, No. 10, (May 2010), pp. 1286–1295

Peppa, M.; Goldberg, T.; Cai, W.; Rayfield, E. & Vlassara, H. (2002). Glycotoxins: a missing link in the "relationship of dietary fat and meat intake in relation to risk of type 2 diabetes in men". *Diabetes Care*, Vol. 25, No. 10, (November, 2002), pp.1898-1899

Patel, S.P. & Katyare S.S. (2006). Insulin-status-dependent modulation of FoF1-ATPase activity in rat liver mitochondria. *Lipids*, Vol. 41, No. 7, (July 2007), pp. 695-703

Petersen, K.F.; Befroy, D.; Dufour, S.; Dziura, J.; Ariyan, C.; Rothman, D.L.; DiPietro, L.; Cline, G.W. & Shulman, G.I. (2003). Mitochondrial dysfunction in the elderly: possible role in insulin resistance. *Science*, Vol. 300, No. 5622, (May 2003), pp. 1140-1142

Pinti, M.; Nasi, M.; Gibellini, L.; Roat, E.; De Biasi, S.; Bertoncelli, L. & Cossarizza A. (2010). The role of mitochondria in HIV infection and its treatment. *J Exp Clin Med*, Vol. 2, No. 4, (August 2010), pp. 145–155

Pijl, H.; Ohashi, S.; Matsuda, M.; Miyazaki, Y.; Mahankali, A.; Kumar, V.; Pipek, R.; Iozzo, P.; Lancaster, J.L.; Cincotta, A.H. & DeFronzo, R.A. (2000). Bromocriptine: a novel approach to the treatment of type 2 diabetes. *Diabetes Care*, Vol. 23, No. 8, (August 2000), pp. 1154-1161.

Price, D.L.; Rhett, P.M.; Thorpe, S.R. & Baynes, J.W. (2001). Chelating activity of advanced glycation end-product inhibitors. *J Biol Chem*, Vol. 276, No. 52, (December 2001), pp. 48967-48972

Ravelojaona, V.; Peterszegi, G.; Molinari, J.; Gesztesi, J.L. & Robert, L. (2007). Demonstration of the cytotoxic effect of Advanced Glycation Endproducts (AGE-s). *J Soc Biol*, Vol. 201, No. 2, n.d., pp.185-188

Rolo, A.P. & Palmeira, C.M. (2006). Diabetes and mitochondrial function: Role of hyperglycemia and oxidative stress. *Toxicol Appl Pharm*, Vol. 212, No. 2, (April 2006), pp. 167-178

Sell, D.R.; Lane, M.A.; Johnson, W.A.; Masoro, E.J.; Mock, O.B.; Reiser, K.M.; Fogarty, J.F.; Cutler, R.G.; Ingram, D.K.; Roth, G.S. & Monnier, V.M. (1996). Longevity and the genetic determination of collagen glycoxidation kinetics in mammalian senescence. *Proc Natl Acad Sci USA*, Vol. 93, No. 1, (January 1996), pp. 485-490

Shen, G X. (2010). Oxidative stress and diabetic cardiovascular disorders: roles of mitochondria and NADPH oxidase. *Can J Physiol Pharmacol*,Vol. 88, No. 3, (March, 2010), pp. 241-248

Singh, R.; Barden, A.; Mori, T. & Beilin, L. (2001). Advanced glycation end-products: a review. *Diabetologia*, Vol. 44, No. 2, (February 2001), pp.129-146

Srinivasan, K. & Ramarao, P. (2007). Animal models in type 2 diabetes research: an overview. *Indian J Med Res*, Vol. 125, No. 3, (March 2007), pp.451-472

Szewczyk, A. & Wojtczak, L. (2002). Mitochondria as a pharmacological target. *Pharmacol Rev*, Vol. 54, No. 1, n.d., pp. 101-127

Tessier, F.J. & Birlouez-Aragon, I. (2010). Health effects of dietary Maillard reaction products: the results of ICARE and other studies. *Amino Acids*, (October 2010), DOI: 10.1007/s00726-010−0776-z

Thornalley, P.J. (1998). Cell activation by glycated proteins. AGE receptors, receptor recognition factors and functional classification of AGEs. *Cell Mol Biol, Vol.* 44, No. 7, (November, 1998), pp. 1013-1023

Turner, R.C.; Cull, C.A.; Frighi, V. & Holman. R.R. (1999). Glycemic control with diet, sulfonylurea, metformin, or insulin in patients with type 2 diabetes mellitus: progressive requirement for multiple therapies (UKPDS 49). UK Prospective Diabetes Study (UKPDS) Group. *JAMA, Vol.* 281, No. 21, (June, 1999), pp.2005-2012

Urios, P.; Grigorova-Borsos, A.M.; Peyroux, J. & Sternberg, M. (2007). Inhibition of advanced glycation by flavonoids. A nutritional implication for preventing diabetes complications?. *J Soc Biol, Vol.* 201, No. 2, n.d., pp. 189-198

Vlassara, H. & Palace, M.R. (2002). Diabetes and advanced glycation endproducts. *J Intern Med, Vol.* 251, No. 2 (February 2002), pp. 87-101

Vojtassak, J.; Blasko, M.; Danisovic, L. Sr.; Carsky, J.; Durikova, M.; Repiska, V.; Waczulikova, I. & Bohmer, D. (2008). In vitro evaluation of the cytotoxicity and genotoxicity of resorcylidene aminoguanidine in human diploid cells B-HNF-1. *Folia Biol, Vol.* 54, No. 4, n.d., pp.109-114

Voziyan, P.A.; Khalifah, R.G.; Thibaudeau, C.; Yildiz, A.; Jacob, J.; Serianni, A.S. & Hudson, B.G. (2003). Modification of proteins in vitro by physiological levels of glucose: pyridoxamine inhibits conversion of Amadori intermediate to advanced glycation end-products through binding of redox metal ions. *J Biol Chem, Vol.* 278, No. 47, (November 2003), pp. 46616-46624

Voziyan, P. A. & Hudson, B.G. (2005). Pyridoxamine as a multifunctional pharmaceutical: targeting pathogenic glycation and oxidative damage. *Cell Mol Life Sci, Vol.* 62, No.15, (August 2005), pp.1671-1681

Waczulikova, I.; Sikurova, L.; Bryszewska, M.; Rekawiecka, K.; Carsky, J. & Ulicna, O. (2000). Impaired erythrocyte transmembrane potential in diabetes mellitus and its possible improvement by resorcylidene aminoguanidine. *Bioelectrochemistry. Vol.* 52, No. 2, (December 2000), pp. 251-256

Waldbaum, S. & Patel, M. (2009). Mitochondria, oxidative stress, and temporal lobe epilepsy. *Epilepsy Res, Vol.* 88, No. 1, (January 2010), pp. 23-45

Watala, C.; Dobaczewski, M.; Kazmierczak, P.; Gebicki, J.; Nocun, M.; Zitnanova, I.; Ulicna, O.; Durackova, Z.; Waczulikova, I.; Carsky, J. & Chlopicki, S. (2009). Resorcylidene aminoguanidine induces antithrombotic action that is not dependent on its antiglycation activity. *Vascul Pharmacol, Vol.* 51, No. 4, (October 2009), pp. 275-283

Watkins, P.B. & Whitcomb, R.W. (1998). Hepatic dysfunction associated with troglitazone. *N Engl J Med, Vol.* 338, No. 13, (March 1998), pp. 916-917

Westwood, M.E.; Argirov, O.K.; Abordo, E.A.; & Thornalley, P.J. (1997). Methylglyoxal-modified arginine residues--a signal for receptor-mediated endocytosis and degradation of proteins by monocytic THP-1 cells. *Biochim Biophys Acta Vol.* 1356, No. 1, (March 1997), pp. 84-94, 1997

Wierusz-Wysocka, B.; Wysocki, H.; Byks, H.; Zozulinska, D.; Wykretowicz, A. & Kazmierczak, M. (1995). Metabolic control quality and free radical activity in diabetic patients. *Diabetes Res Clin Pract, Vol.* 27, No. 3, (March, 1995), pp.193-197

Wirth, D.D.; Baertschi, S.W.; Johnson, R.A.; Maple, S.R.; . Miller, M.S.; Hallenbeck, D.K. & Gregg, S.M. (1998). Maillard reaction of lactose and fluoxetine hydrochloride, a secondary amine. *J Pharm Sci, Vol.* 87, No. 1, (January 1998), pp.31-39

Yamamoto, M.; Ozawa, K. & Tobe T. (1981). Roles of high blood glucose concentration during hemorrhagic shock in alloxan diabetic rats. *Circ Shock,* Vol. 8, No. 1, n.d., pp. 49-57

6

Medical Plant and Human Health

Ahmed Morsy Ahmed

*Faculty of Agriculture,
Ain shams University,
Egypt*

1. Introduction

1.1 Primary healthcare and traditional medicine

Primary Health Care (PHC) is the key to the development of a national health policy based on practical, scientifically sound and socially acceptable methods and technology made universally acceptable to individuals and families in the community and through their full participation and at a cost that the community and the country can afford, in order to maintain, at every stage of their development, in the spirit of self-reliance and self determination. It is the first level of contact for the individual, family and the community within the national health care system, bringing health care as close as possible to where people live and work and thus constitutes the first element of a continuing health care process (WHO, 1978a). A health system, based on primary health care was adopted as the means of achieving the goal of health for all. Most developing countries of the world, for which the scheme was designed, have failed to seriously implement it up till this moment.

Traditional Medicine is defined by the World Health Organization (WHO, 1978a) as the sum total of knowledge or practices whether explicable or inexplicable, used in diagnosing, preventing or eliminating a physical, mental or social disease which may rely exclusively on past experience or observations handed down from generation to generation, verbally or in writing. It also comprises therapeutic practices that have been in existence often for hundreds of years before the development of modern scientific medicine and are still in use today without any documented evidence of adverse effects. The explicable form of Traditional Medicine can be described as the simplified, scientific and the direct application of plant, animal or mineral materials for healing purposes and which can be investigated, rationalized and explained scientifically.

1.2 Herbal medicine

The use of *Salix alba*, the willow plant (containing the salicylates) for fever and pains which led to the discovery of aspirin, would belong to this form of Traditional Medicine. Herbal medicines, which squarely belong to this form, are regarded by the World Health Organization, as finished and labeled medicinal products that contain, as active ingredients, aerial or underground parts of identified and proven plant materials, or combination thereof, whether in crude form or as plant preparations. They also include plant juices, gums, fatty oils, essential oils etc (WHO, 1978a). There are several other official modern drugs today, which were originally developed like aspirin through traditional medicine e.g.

morphine, digoxin, quinine, ergometrine, reserpine, atropine, etc and all of which are currently being used by orthodox medicine in modern hospitals all over the world.

The inexplicable form of Traditional Medicine on the other hand, is the spiritual, supernatural, magical, occultic, mystical, or metaphysical form that cannot be easily investigated, rationalized or explained scientifically e.g. the use of incantations for healing purposes or oracular consultation in diagnosis and treatment of diseases. The explanation is beyond the ordinary scientific human intelligence or intellectual comprehension. Plants are reputed in the indigenous systems of medicine for the treatment of various diseases (Arise *et al.*, 2009). Phyto-chemicals isolated from plant sources are used for the prevention and treatment of several medical problems including diabetes mellitus (Waltner- Law *et al.*, 2002). There are more than 800 plant species showing a hypoglycemic activity. The World Health Organization (1980) has also recommended the evaluation of the effectiveness of plants in conditions where safe modern drugs are lacked.

2. Major groups of antimicrobial compounds from plants

Plants have an almost limitless ability to synthesize aromatic substances, most of which are phenols or their oxygen-substituted derivatives (Geissman 1963). Most are secondary metaboltes, of which at least 12,000 have been isolated, a number estimated to be less than 10% of the total (Schultes 178). In many cases, these substances serve as plant defense mechanisms against predation by microorganisms, insects, and herbivores. Some, such as terpenoids, give plants their odors; others (quinones and tannins) are responsible for plant pigment. Many compounds are responsible for plant flavor (e.g., the terpenoid capsaicin from chili peppers), and some of the same herbs and spices used by humans to season food yield useful medicinal compounds.

Simple phenols and phenolic acids, The mechanisms thought to be responsible for phenolic toxicity to microorganisms include enzyme inhibition by the oxidized compounds, possibly through reaction with sulfhydryl groups or through more nonspecific interactions with the proteins (Mason and Wasserman 1987).Phenolic compounds possessing a C_3 side chain at a lower level of oxidation and containing no oxygen are classified as essential oils and often cited as antimicrobial as well. Eugenol is a well-characterized representative found in clove oil. Eugenol is considered bacteriostatic against both fungi (Duke 1985) and bacteria (Thomson 1978). Quinones are aromatic rings with two ketone substitutions. They are ubiquitous in nature and are characteristically highly reactive. Probable targets in the microbial cell are surface-exposed adhesins, cell wall polypeptides, and membrane-bound enzymes. Quinones may also render substrates unavailable to the microorganism. As with all plant-derived antimicrobials, the possible toxic effects of quinones must be thoroughly examined.Kazmi et al.

Flavones, flavonoids, and flavonols activity is probably due to their ability to complex with extracellular and soluble proteins and to complex with bacterial cell walls, as described above for quinones. More lipophilic flavonoids may also disrupt microbial membranes (Tsuchiya 1996). Tannins are found in almost every plant part: bark, wood, leaves, fruits, and roots (Scalbert 1991). This group of compounds has received a great deal of attention in recent years, since it was suggested that the consumption of tannin-containing beverages, especially green teas and red wines, can cure or prevent a variety of ills (Serafini et al 1994) their mode of antimicrobial action, may be related to their ability to inactivate microbial adhesins, enzymes, cell envelope transport proteins, etc. They also complex with polysaccharide (Ya et al 1988). The antimicrobial significance of this particular activity has not been explored. There is also evidence for direct inactivation of microorganisms: low tannin concentrations modify the morphology of germ tubes of *Crinipellis perniciosa* (Brownlee et al 1990). Tannins in plants inhibit insect growth (Schultz 1988) and disrupt digestive events in ruminal animals (Butler 1988).

Terpenoids and Essential Oils are secondary metabolites that are highly enriched in compounds based on an isoprene structure. They are called terpenes, their general chemical structure is $C_{10}H_{16}$, and they occur as diterpenes, triterpenes, and tetraterpenes (C_{20}, C_{30}, and C_{40}), as well as hemiterpenes (C_5) and sesquiterpenes (C_{15}). When the compounds contain additional elements, usually oxygen, they are termed terpenoids. The mechanism of action of terpenes is not fully understood but is speculated to involve membrane disruption by the lipophilic compounds. Accordingly, Mendoza et al. (Mendoza et al 1997) found that increasing the hydrophilicity of kaurene diterpenoids by addition of a methyl group drastically reduced their antimicrobial activity.

Alkaloids are Heterocyclic nitrogen compounds. Alkaloids have been found to have microbiocidal effects (including against *Giardia* and *Entamoeba* species Ghosha et al 1996, the major antidiarrheal effect is probably due to their effects on transit time in the small intestine. Berberine is an important representative of the alkaloid group. It is potentially effective against trypanosomes (Freiburghaus et al 1996) and plasmodia Omulokoli et al 1997). The mechanism of action of highly aromatic planar quaternary alkaloids such as berberine and harmane (Hopp 1976) is attributed to their ability to intercalate with DNA (Phillipson and O'Neill 1987). The mechanism of lectins and polypeptides may be back to the formation of ion channels in the microbial membrane (Terras et al. 1993) or competitive inhibition of adhesion of microbial proteins to host polysaccharide receptors (Sharon and Ofek 1986). Recent interest has been focused mostly on studying anti-HIV peptides and lectins, but the inhibition of bacteria and fungi by these macromolecules, such as that from the herbaceous *Amaranthus*, has long been known (De Bolle et al 1996).

2.1 Ethnobotanical study

Out of around 1076 species recorded so far from Semifinal Biosphere Reserve, more than 200 species are attributed with medicinal uses (Rout, 2004). This system of using herbs and different biological active ingredients in treating various diseases had become a part of their culture till recent years. Entrance of market economy gave rise to exploitation of natural resources and thereby depleting our resources base. The most affected part in this process was medicinal plants, which is most sensitive and delicate in the environment of forest. These medicinal plants gain further importance in the region where modern medical health facilities are either not available or not easily accessible. Although our ancient sages through hit and trial method developed herbal medicines, the reported uses of plant species do not certify efficacy.

The present preliminary report on the uses of some plant species need to pharmacologically screened, chemically analyzed and tested for bioactive activities. Pharmacological screening of plant extracts provides insight to both their therapeutic and toxic properties and helps in eliminating the medicinal plants or practices that may be harmful. The plant parts used for medical preparation were bark, flowers, rhizomes, roots, leaves, seeds, gum and whole plants. In some cases the whole plant including roots was utilized. The use of plant derived products containing high of dietary fiber and complex polysaccharide for the management of diabetes have been proposed (Jenkins et al., 1976). Natural products especially of plant origin have been found to be potential sources of novel molecules for the treatment of diabetes (Farnsworth, 1994; Marles and Farnsworth, 1995).

Considering the rate at which the vegetation is getting depleted in this part of the world, therefore it is needed to document the precious knowledge of these plants and to search for more plants with antidiabetic potential. The search for anti-diabetic agents has been focused on plants because of their ready availability, effectiveness, affordability and probably due to low side effects (Marles and Farnsworth, 1995). Ethnobotanical study has been the method often used to search for locally important plant species with low side effects especially for the discovery of crude drugs (Farnsworth, 1994). The present study therefore is a documentation of plants and plant parts used for the management of diabetes mellitus by traditional healers of the area (Table 1).

Diagnosis methods	Respondent (%)
Signs and symptoms	
Loss of body weight	100
Body weakness	100
Excessive urination	100
Presence of sugar in urine	100
Excessive thirsty	53
Duration of treatment	
Short duration	40
Long duration	60
Efficacy of plant treatment on patients	
Disappearance of sugar in urine 100	100
Reduction in body weakness 100	100
Normal body weight 100	100
Reduction in frequency of urination	100
Traditional healers claim of no adverse effect after treatment	
Yes	100
No	0
Traditional healers claim of total cure after treatment	
Yes	80
No	20

Table 1. Diagnosis methods of diabetes mellitus by the herbalists using herbal preparation

2.2 Diabetes mellitus
2.2.1 What is Diabetes?

The word "diabetes" (a Greek word that means "to pass through") was first used by Aretaeous of Capadocia in the 2nd century AD to describe a condition that is characterized by excess of sugar in blood and urine, hunger and thirst (MacFrlance et al.,1997) and the adjective "mellitus" (a latin-greek word that means "honey") was introduced by the English physician John Rollo so as to distinguish the conditions from other polyuric diseases, in which glycosuria does not occur (Rollo 1797). People suffering from diabetes are not able to produce or properly use insulin in the body and therefore chronic hyperglycemia occurs. In addition, the diabetic individual is prone to late onset complications (Fujisawa et al., 2004), such as retinopathy, neuropathy and vascular diseases, that are largely responsible for the morbidity and mortality observed in diabetic patients.

Diabetes mellitus is a chronic metabolic disorder characterized by widespread complications. It is the world's largest endocrine disease associated with increased morbidity and mortality rates (Ghosh and Surawanshi, 2001). The chronic hyperglycemia of diabetes is associated with long-term damage, dysfunction and failure of various organs (Lyra *et al.*, 2006). Liver involvement is one of the leading causes of death in diabetes mellitus. The mortality rate from the hepatic affection is greater than that from the cardiovascular complications. The spectrum of liver implication in diabetes ranges from non-alcoholic fatty liver disease to cirrhosis and eventually hepato-cellular carcinoma (Keith *et al.*, 2004). Liver, an insulin-dependent organ, plays a pivotal role in glucose and lipid homeostasis. It participates in the uptake, oxidation and metabolic conversion of free fatty acids and in the synthesis of cholesterol, phospholipids and triglycerides.

Several mechanisms are implicated in the pathogenesis of the functional and morphological alterations of the liver of diabetic patients (Brixova, 1981; Moller, 2001). There are two main types of diabetes, namely type I and type II (World Health Organization. Definition, Diagnosis and Classification of Diabetes mellitus and its omplications. Part 1: Diagnosis and Classification of Diabetes Mellitus (Department of Non communicable Disease Surveillance, Geneva, 1999)). Type I Diabetes, that is called insulin-dependent diabetes mellitus (IDDM) or juvenile onset diabetes develops when the body's immune system destroys pancreatic β-ells, the only cells in the body that produce the hormone insulin that regulates blood glucose. This type of diabetes usually strikes children and adults and the need for insulin administration is determinant for survival. Type I diabetes accounts for 5% to 10% of all diagnosed cases of diabetes and the risk factors may be autoimmune, genetic, or environmental.

On the other hand, type II diabetes, also called non–insulin-dependent diabetes mellitus (NIDDM) or adult-onset diabetes, accounts for about 90% to 95% of all diagnosed cases of diabetes. It usually begins as insulin resistance, a disorder in which the cells do not use insulin properly and as the need for insulin rises; the pancreas gradually loses its ability to produce it. This type of diabetes is associated with older age, obesity, family history of diabetes, history of gestational diabetes, impaired glucose metabolism, physical inactivity, and race/ethnicity. It must be noted thought that in the last decay type II diabetes in children and adolescents is being diagnosed more frequently (Fagot-Gampagna & Narayan 2001). In the case of the IDDM, insulin is of crucial importance for the survival of the patients. On the other hand, in the case of NIDDM the treatment includes medicines, diets and physical training.

2.2.2 Ant diabetic medicines: Important medicinal plants used in diabetes treatment; important active

2.2.2.1 Compounds involved in diabetes mellitus treatment

Many kinds of antidiabetic medicines have been developed for the patients and most of them are chemical or biochemical agents aiming at controlling or/and lowering blood glucose to a normal level. Despite the impressive advances in health sciences and medical care, there are many patients who are using alternative therapies alone or complementary to the prescribed medication. Traditional plant remedies or herbal formulations exist from ancient times and are still widely used, despite all the controversy concerning their efficacy and safety (Huxtable 1990; Fugh-Berman 2000), to treat hypoglycemic and hyperglycemic conditions all over the world.

It must be noted that many ethno-botanical surveys on medicinal plants used by the local population have been performed in different parts of the world and there is a considerable number of plants described as antidiabetic. In addition a variety of compounds have been isolated (alkaloids, glycosides, terpenes, flavonoids, etc) but further studies need to be done so as these 'leads' to develop into clinically useful medicines. To date, met form in (a biguanide) is the only drug approved for treatment of type II diabetes mellitus. It is a derivative of an active natural product, galegine, isolated from the plant Galega officinalis L. (Witters 2001). The followinf table illustrate some of these plant involved in diabetes mellitus treatment

Scientific name	Family	Active constituent	Part Used	Folk Medical Uses
Abelmoschus moschatus	Malvaceae	myricetin	aerial part	hypoglycemic action decrease the plasma glucose concentrations myricetin has an ability to enhance glucose utilization to lower plasma glucose in diabetic rats with deficient insulin levels.
Azadirachta indica	Meliaceae		dried powder of root and leaves	caused significant lowering of blood sugar and reduction in serum lipids
Cornus macrophylla	Cornaceae			prevent and treat diabetic complications
Achyranthes aspera L.	Amaranthaceae		powdered whole plant	hypoglycemic effect
Achyrocline satureioides	Asteraceae	achyrofuran	powdered whole plant	lowered blood glucose levels
Acosmium panamense	Fabaceae		water and butanolic extracts	lowered the plasma glucose levels
Aegle marmelose	Rutaceae		leaf extract	similar hypoglycemic activity to that of insulin treatment Treatment with the leaf extract showed improved functional state of pancreatic β cells. The results indicate the potential hypoglycemic effect of the leaf extract, possibly involved in processes for the regeneration of damaged pancreas.

Table 2. Important medicinal Plants used in Diabetes treatment.

Scientific name	Family	Active constituent	Part Used	Folk Medical Uses
Allium sativum L	(Liliaceae	allicine		decreased the concentration of serum lipids, blood glucose and activities of serum enzymes like alkaline phosphatase, acid phosphatase and lactate dehydrogenase and liver G6Pase
Allium cepa L.	(Liliaceae	S-methyl cysteine sulfoxide		lowered the levels of malondialdehyde, hydroperoxide and conjugated dienes in tissues exhibiting antioxidant effect
		(SMCS)		on lipid peroxidation in experimental diabetes The hypoglycemic and hypolipidaemic actions of Allium cepa were associated with antioxidant activity, since onion decreased SOD activities while no increased lipid hydroperoxide and lipoperoxide concentrations were observed in diabetic rats treated with Allium cepa
Aloe vera	(Liliaceae		leaf pulp and gel extracts	contain a hypoglycemic agent which lowers the blood glucose levels showed hypoglycemic activity on type I and II diabetic maintained the glucose homeostasis by controlling the carbohydrate metabolizing enzymes (Rajasekaran et al., 2004).
Aloe barbadensis	Liliaceae		leaves	the bitter principle may be mediated through stimulating synthesis and/or release of insulin from the β cells of Langerhans
Andrographis paniculata	Acanthaceae	andrographolide	crude ethanolic extract	antidiabetic effect may be attributed, at least in part, to increased glucose metabolism andrographolide can increase the glucose utilization to lower plasma glucose in diabetic rats lacking insulin.

Table 2. Important medicinal Plants used in Diabetes treatment. (continuation)

Scientific name	Family	Active constituent	Part Used	Folk Medical Uses
Angylocalyx pynaertii	Leguminosae	sugar-mimic alkaloids	pod extract	specific inhibitors of alpha-L fucosidase with no significant inhibitory activity towards other glycosidases
Areca catechu		arecolin		Arecoline have hypoglycemic activity in animal model of diabetes upon s.c. administration
Averrhoa bilimbi L.	Oxalidaceae		ethanolic extract	ethanolic extract has hypoglycemic, hypotriglyceridemic, anti-lipid peroxidative and anti-atherogenic properties in STZ-diabetic rats.
Bauhinia forficata L.	Leguminosae	kaempfer itr in	n-butanol extract	blood glucose-lowering effect in normal and diabetic rats
Beta vulgaris L. var. cicla	Chenopodiaceae	Betavulgaroside I, II, III and IV	roots	hypoglycemic effects that was demonstrated by a per os glucose tolerance test in rats after their per os administration
Bidens pilosa	Asteraceae	3-beta-D-glucopyranosylo xy-1-hydroxy-6(E)-tetradecene-8,10,12-triyne	Butanol extract of whole plant	inhibited the differentiation of naive helper T cells (Th0) into Th1 cells but enhanced their transition into type II helper T (Th2) cells using an in vitro T cell differentiation assay
Bryonia alba L.	Curcubiaceae	derivatives of trihydroxyoctade cadienoic acid	roots	Restore the disordered lipid metabolism of alloxan-diabetic rats
Bumelia sartorum	Sapotaceae	bassic acid	ethanol extract of root bark	altered glucose tolerance in alloxan-induced diabetic rats, enhanced glucose uptake in skeletal muscle and significantly inhibited glycogenolysis in the liver.
Caesalpinia ferrea Mart	Leguminosae	ellagic acid and 2'-(2,3,6-trihydroxy-4-carboxyphenyl) ellagic aci		might contribute to the relief of the long-term diabetic complications
Camellia sinensis	Theaceae	epigallocatechin-3-O-gallate		may be beneficial in the prevention of diabetes mellitus
Cassia tora	Fabaceae	a-tocopherol, ascorbic acid and maltodextrin	soluble fiber extracted	improve serum lipid levels in type II diabetic subjects without serious adverse effects

Table 2. Important medicinal Plants used in Diabetes treatment.(continuation)

Scientific name	Family	Active constituent	Part Used	Folk Medical Uses
Cecropia obtusifolia Bertol	Cecropiaceae	isoorientin and 3-caffeoylquinic acid	water and butanolic extracts prepared from leaves	beneficial effects on carbohydrate and lipid metabolisms when it was administered as an adjunct on patients with type II diabetes with poor response to conventional medical treatment.
Cleome droserifolia (Capparidaceae)				possessed a postprandial hypoglycemic effect but also suppressed the hepatic glucose release in the fasting state in a comparable way to this of insulin. The plant also possessed hypo-cholesterolemic effect, most pronounced on the LDL cholesterol.
Cnidium officinale Makino	Apiaceae			inhibited the high glucoseinduced proliferation of GMCs partially through TGF-beta1 production
Coccinia indica	Cucurbitaceae		ethanolic extract of Coccinia indica leaves	normalized blood glucose and caused marked improvement of altered carbohydrate metabolizing enzymes during diabetes.
Commelina communis	Commeliaceae	pyrrolidine alkaloid, 2,5-dihydroxymethyl-3,4-dihydroxypyrrolidine and four piperidine alkaloids, 1-deoxymannojirimycin, 1-deoxynojirimycin, alpha-homonojirimycin and 7-O-beta-D-glucopyranosyl alpha-homonojirimycin	methanolic extract	inhibitoryactivity against alpha-glucosidase

Table 2. Important medicinal Plants used in Diabetes treatment. (continuation)

Scientific name	Family	Active constituent	Part Used	Folk Medical Uses
Croton cajucara	Euphorbiaceae	trans-hydrocortin		The mentioned compound also effectively lowered the blood sugar levels in glucose fed normal rats.
Cryptolepis sanguinolenta	Asclepiadaceae	cryptolepine		antihyperglycemic effect of cryptolepine leaded to a significant decline in plasma insulin concentration, associated with evidence of an enhancement in insulin mediated glucose disposal. Finally, cryptolepine increased glucose uptake by 3T3-L1 cells.
Dioscorea dumetorum	Dioscoreaceae	dioscoretine	methanol extrac	normal and alloxan diabetic rabbits produced significant hypoglycemic effects and the hypoglycemic effects were compared to those of to butamide (Iwu et al., 1998)
Galega officinalis L.	Leguminosae	Metformin		Treat symptoms now ascribed to type II diabetes
Gentiana olivieri Griseb	Gentianaceae	isoorientin		Isoorientin exhibited significant hypoglycemic and antihyperlipidemic effects
Gymnema sylvestre	Asclepiadaceae	gymnemic acid I	Water soluble extracts the leaves	Gymnemic acids II and III showed potent inhibitory activity on glucose uptake.
Hintonia latiflora	Rubiaceae	coutaraegenin		coutareagenin, one of the active substances contained in the Hintonia latiflora bark, produces a reduction of the diabetic elevated blood sugar levels
Hydnocarpus wightiana	Arcariaceae	Hydnocarpin, Luteolin, And isohydnocarpin	acetone extract of the seed hulls	the presence of antioxidant molecules along with their enzyme inhibitory activities in the acetone extract of Hydnocarpus wightiana seed hulls may be responsible for the antidiabetic Properties

Table 2. Important medicinal Plants used in Diabetes treatment. (continuation)

Scientific name	Family	Active constituent	Part Used	Folk Medical Uses
Lagerstroemia speciosa L.	Lythraceae	Lagerstroemin and flosin B	hot-water extract from banaba leaves	effects on controlling of the level of plasma glucose in type II diabetes mellitus.
Larrea tridentata	Zygophyllaceae	masopropol		lower plasma glucose concentration in two mice models of type II diabetes after per os administration (Luo et al., 1998)
Myrcia multiflora	Myrtaceae	myrciacitrin I	ethyl acetate-soluble portion,	show inhibitory activities on ALR2 and alpha-glucosidase as well as on the increase of serum glucose level in sucrose-loaded rats and in alloxan-induced diabetic mice (Yoshikawa et al., 1998).
Paeonia lactiflora Pall.	Panunculaceae	paeoniflorin		the mentioned glucoside reduced the elevation of blood sugar in glucose challenged rats. Increase of glucose utilization by paeoniflorin can thus be considered (Hsu et al., 1997).
Pandanus odorus Ridl.	Pantadaceae	4-hydroxybenzoic acid		4-hydroxybenzoic acid caused a decrease in plasma glucose levels dose-dependently. The compound did not affect serum insulin level and liver glycogen content in the diabetic model, but increased glucose consumption in normal and diabetic rat diaphragms.
Phyllanthus sellowianus	Euphorbiaceae	isoqu ercitrin and rutin		hypoglycemic and diuretic agent reduction in blood glucose levels
Pueraria thunbergiana	Leguminosae	kaikasaponin III		kaikasaponin III (as a mixture of isomers), may exhibit its hypoglycemic and hypolipidemic effects by up-regulating or down-regulating antioxidant mechanisms via the changes in Phase I and II enzyme activities .
Stevia rebaudiana Bertoni	Compositae	stevioside		Stevioside was able to regulate blood glucose levels by enhancing not only insulin secretion, but also insulin utilization in insulin-deficient rats; the latter was due to decreased PEPCK gene expression in rat liver by stevioside's action of slowing down Gluconeogenesis (Chen et al., 2005).

Table 2. Important medicinal Plants used in Diabetes treatment. (continuation)

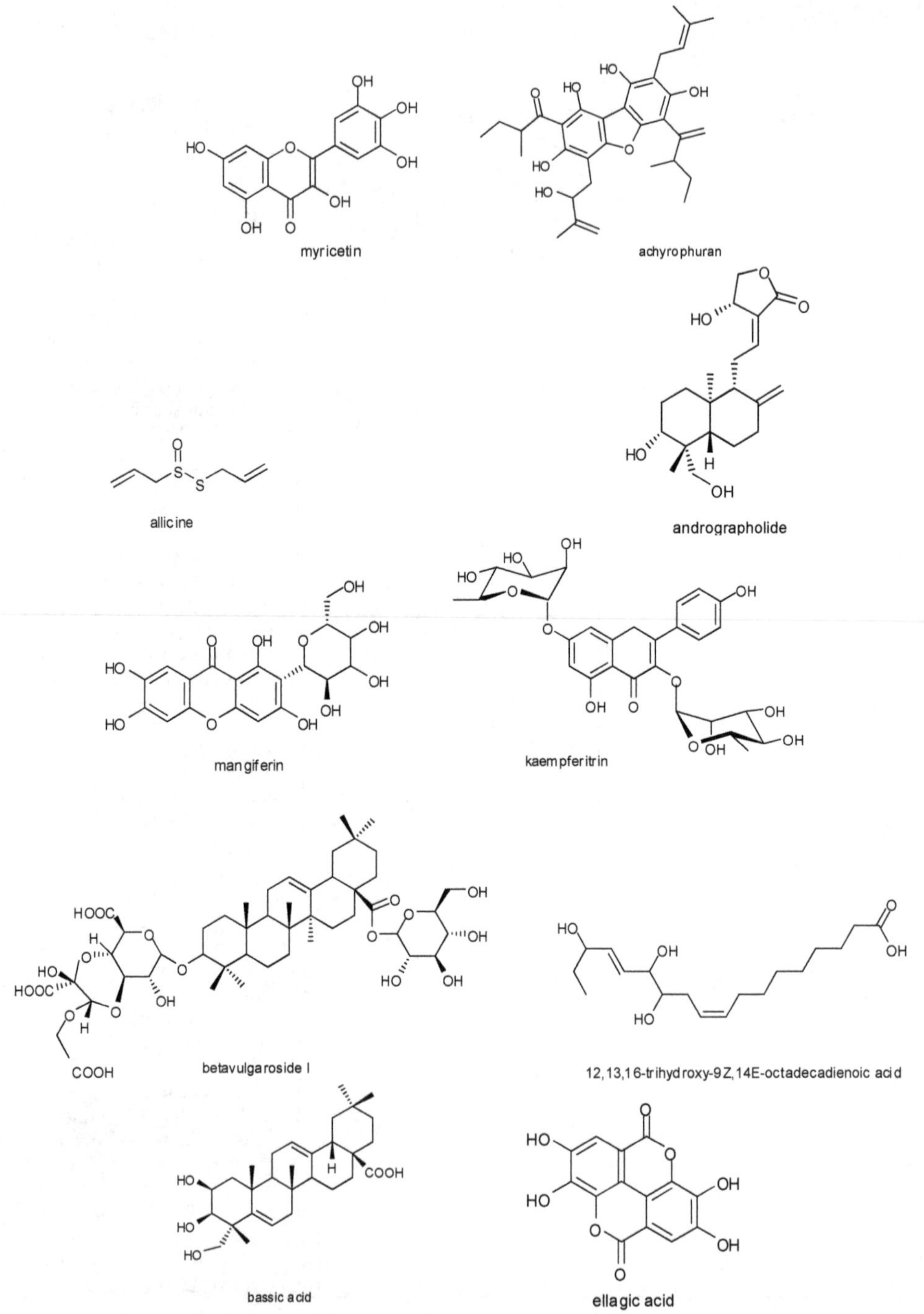

Fig. 1. Some important active compound involved in diabetes mellitus treatment.

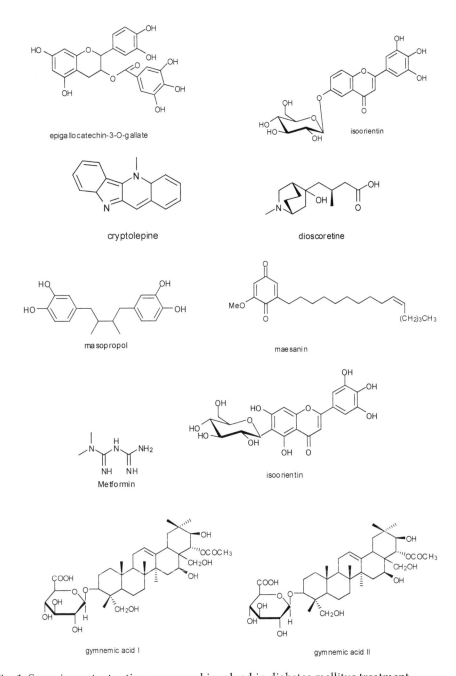

Fig. 1. Some important active compound involved in diabetes mellitus treatment. (continuation)

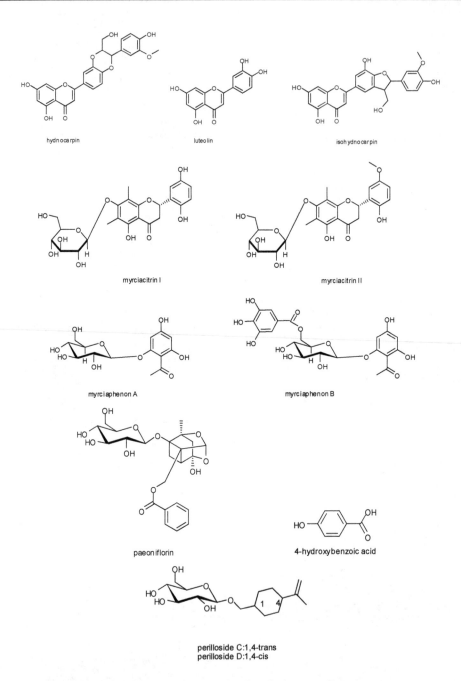

Fig. 1. Some important active compound involved in diabetes mellitus treatment.
(continuation)

3. Some biomedical application for some plant used in treating hypoglycemic and hyperglycemic conditions

A- Amelioration of hyperglycemia in STZ-diabetic rats.

Several approaches were conducted in order to investigate the hypoglycemic and hypolipidemic activities of plant extract on experimental animals (STZ-diabetic rats). The derived data showed that some extracts are potent in the amelioration of hyperglycemia in STZ-diabetic rats and is a potential source for the isolation of new per os active agent(s) for anti-diabetic therapy like what observed in Averrhoa bilimbi ethanolic extract which has hypoglycemic, hypotriglyceridemic, anti-lipid peroxidative and anti-atherogenic properties in STZ-diabetic rats (Purshparaj et al., 2001). Another way was by the inhibition of myeloperoxidase activity and decreased lipid peroxidation, induced by ascorbyl radical either in microsomes or in asolectin and phosphatidylcholine liposomes as what observed in Bauhinia forficate due to the kaemperitrin effect .that showed high reactivity with the free radical 1,1-diphenyl-2-picryl hydrazyl (DPPH),

In the plant *Beta vulgaris L. var. cicla (Chenopodiaceae)* the anti-diabetic effect   in experimental animals may be to the reduction in blood glucose levels by the regeneration of the β cells (Bolkent et al., 2000). Betavulgarosides I, II, III, IV and oleanolic acid oligoglycosides produced hypoglycemic effects that was demonstrated by a per os glucose tolerance test in rats after their per os administration (Yoshikawa et al., 1996). Or through stimulating the synthesis/release of insulin from the β cells of Langerhans like the effect of *Biophytum sensitivum* and/or also mediated through enhance secretion of insulin from the β cells of Langerhans or through an Extra pancreatic mechanism *(Catharanthus roseus L. (Apocynaceae))*. Or by increasing the number of islet β cells and that of islet α cells was decreased in STZ-diabetic rats like what happen in Dendrobium candidum (Orchidaceae), Its mechanism of action probably involves the stimulation of the secretion of insulin from β cells and the inhibition of the secretion of glucagons from α cells and it can probably decrease the decomposition of liver glycogen and increase the synthesis of liver glycogen.

The Administration of *Coccinia indica (Cucurbitaceae)* leaf extract, an indigenous plant used in Ayurvedic medicine in India, to normal and STZ-diabetic animals exhibited significant hypoglycemic and antihyperglycemic effect and reversed the associated with diabetes biochemical alterations (Venkateswaran & Pari 2002). The results indicated that the per os administration of Coccinia indica leaf extract to diabetic animals normalized blood glucose and caused marked improvement of altered carbohydrate metabolizing enzymes during diabetes. The antioxidant effects of an ethanolic extract of Coccinia indica leaves was studied in STZ-diabetic rats (Venkateswaran & Pari 2003). Per os administration of Coccinia indica leaf extract resulted in a significant reduction in thiobarbituric acid reactive substances and hydroperoxides and a significant increase in reduced GSH, SOD, CAT, glutathione peroxidase and glutathione-S-transferase in liver and kidney of STZ-diabetic rats, which clearly showed the extract's antioxidant property. The ethnopharmacological use of *Gongronema latifolium* in ameliorating the oxidative stress found in diabetics and indicating promise of possible use in lessening morbidity in affected individuals.

The obtained result by Kim et al 2006 suggest that the administration of *Chrysanthemum coronarium* and *Morus alba,* have a hypoglycemic effect in diabetic rats and their effect was equivalent to that of glibenclamide. The administration of *C.unshiu* shows more antihyperlipidemic effect than antidiabetic effect. The effect of aqueous extracts from four medicinal plants on the blood glucose levels of experimental animals was determined at various time intervals for 9 h after oral administration at 100 mg dose kg-1 b.wt. (Fig. 2). There was a significant elevation in the blood glucose level by 3.3-5 times during experimental time period in alloxan-induced diabetic rats, when compared to normal rats. The administration of *C. coronarium* extract caused the blood glucose levels of diabetic rats to 83.4, 67.6, 75.1, 81.1 and 74.3% at the time interval of 1, 3, 5, 7 and 9 h, respectively (p<0.05). Maximum reduction of 32.4% was observed 3 h after treatment. The administration of *M. alba* extract produced the most significant reduction (p<0.05) among four medicinal plants in the blood glucose levels of 34, 41, 33 and 35% at 3, 5, 7 and 9 h respectively.

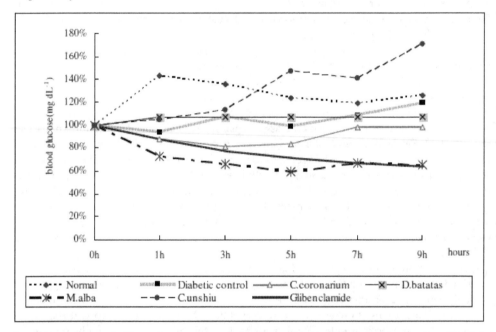

Fig. 2. Percentage of effect of experimental plants on plasma glucose levels compared with 0 h data in alloxaninduced diabetic rats (%). Values are mean percent of blood glucose concentration (n= 6). According to the study obtained by Kim et al., 2006.

B- Hypoglycemic, hypolipidemic hypotriglyceridemic, anti-lipid peroxidative and anti-atherogenic properties of these agents in different medicinal plant:

Insulin resistance (hyperinsulinemia) is now recognized as a major contributor to the development of glucose intolerance, dyslipidemia and hypertension in type II diabetic patients. Aqueous extracts of Pterocarpus marsupium bark (Fbaceae) and Trigonella foenum-graecum. (Leguminosae) seeds have been shown to exert hypoglycemic/antihyperglycemic effect in experimental as well as in clinical settings (Grover et al., 2005). Results of this study, in addition to previous clinical benefits of

Pterocarpus marsupium seen in type II diabetic subjects, are suggestive of usefulness of its bark in insulin resistance, the associated disorder of type II diabetes. Though several antidiabetic principles (epicatechin, pterosupin, marsupin and pterostilbene) have been identified.

The administration of trihydroxyoctadecadienoic acid and its derivatives obtained from the roots of the native Armenian plant *Bryonia alba (Curcubiaceae)* was found to restore the disordered lipid metabolism of alloxan-diabetic rats (Karageuzyan et al., 1998). These derivatives of trihydroxyoctadecadienoic acid can correct major metabolic abnormalities typical of severe diabetes mellitus they can influence the profile of the formation of stable prostaglandins by actions such as downstream of prostaglandin endoperoxides. In *Camellia sinensis (Theaceae)* species, the blood glucose lowering activity of Camellia sinensis was studied by many workers. It has recently been reported that the major green tea polyphenolic constituent, epigallocatechin 3- gallate, mimics the cellular effects of insulin including the reductive effect on the gene expression of rate-limiting gluconeogenic enzymes in a cell culture system (Koyama et al., 2004). Per os administration of green tea that contains this polyphenolic constituent caused a reduction in the levels of mRNAs for gluconeogenic enzymes, PEPCK and G6Pase in the mouse liver. Epigallocatechin 3-gallate alone was also found to down-regulate the gene expression of these enzymes in vitro.

Bumelia sartorum (Sapotaceae), the hypoglycemic effect may be similar to chlorpropamide and possibly due to an enhanced secretion of insulin from the islets of Langerhans or an increased utilization of glucose by peripheral tissues. Besides hypoglycemic activity, the ethanol extract also elicited significant anti-inflammatory activity, but did not show any significant effects on blood pressure, respiration or on the various isolated tissue preparations studied. Bassic acid, an unsaturated triterpene acid isolated from an ethanol extract of Bumelia sartorum root bark, elicited significant hypoglycemic activity in alloxan-diabetic rats and altered the pattern of glucose tolerance in these animals when administered per os (Naik et al., 1991). In addition, bassic acid treatment increased significantly the glucose uptake process and glycogen synthesis in isolated rat diaphragm. Bassic acid treatment increased plasma insulin levels significantly in alloxandiabetic rats. It was therefore suggested that the hypoglycemia activity of bassic acid may be mediated through enhanced secretion of insulin from the pancreatic β cells.

Compelling evidence of *Bidens pilosa (Asteraceae)* suggests that infiltrating CD4 type I helper T (Th1) cells in the pancreatic islets play a pivotal role in the progression of diabetes in non-obese diabetic mice. Treatment with a butanol fraction of Bidens blood glucose and insulin in non-obese diabetic mice in a dose-dependent manner and elevated the serum IgE levels regulated by Th2 cytokines (Chang et al., 2004). Moreover, the butanol fraction inhibited the differentiation of naive helper T cells (Th0) into Th1 cells but enhanced their transition into type II helper T (Th2) cells using an in vitro T cell differentiation assay. The butanol fraction of *Biderns pilosa* and its polyacetylenes can prevent diabetes possibly via suppressing the differentiation of Th0 cells into Th1 cells and promoting that of Th0 cells into Th2 cells.

The results of a reported study of *Gongronema latifolium (Asclepiadaceae)* suggest that the extracts from leaves could exert their antidiabetic activities through their antioxidant properties (Ugochukwu & Bababy 2002). The aqueous and ethanolic extracts were tested in order to evaluate their effect on renal oxidative stress and lipid peroxidation in non-diabetic and STZ-diabetic rats after per os administration (Ugochukwu & Makini 2003). The ethanolic extract appeared to be more effective in reducing oxidative stress, lipid peroxidation, and increasing the GSH/GSSG ratio, The antihyperglycemic effects of aqueous and ethanolic extracts from Gongronema latifolium leaves was also investigated on glucose and glycogen metabolism in liver of non-diabetic and STZ-diabetic rats (Ugochukwu & Bababy 2003). The data showed that the ethanolic extract from the plant's leaves had antihyperglycemic potency, which was suggested to be mediated through the activation of HK, PFK, G6PD and inhibition of GK in the liver.

According to the study occurred by Mohammad Khalil, et al. (2010), it is found found that, the treatment of the diabetic rats with *Citrullus colocynthis* pulp extract, in the livers showed, more or less, an improvement in the histological architecture with persistence of the cytoplasmic vacuoles in some hepatocytes that could be attributed to the residual adverse effect of the diabetic affliction. But the noticed apparent general improvement signifies that *Citrullus colocynthis* could possess cyto-protective abilities on the hepatocytes. The present findings are supported by those announced by Aburjai *et al.* (2007) who confirmed the anti-diabetic properties of *Citrullus colocynthis* extract. Also, Sebbagh *et al.* (2009) stated that this plant could improve the streptozotosin-induced diabetes in rats. Bujanda *et al.* (2008) owed the similar steatotic inhibitory effect of resveratol to inhibition of the tumor necrosis factor alpha production, lipid per-oxidation and oxidative stress.

The damaged sinusoids, the aemorrhage and the inflammatory cell infiltration encountered in the liver of the diabetic animals, might be due to the hyperglycemic state. Seifalian *et al.* (1999), in an analogous study in rabbits, stressed on that the sinusoidal affection is correlated with the severity of fat accumulation in the parenchymal cells. According to the findings of Khan and Chakarabarti (2003); Hayden *et al.* (2005) and Ban and Twigg (2008), hyperglycemia is the main offending factor in the onset of the micro-vascular diabetic complications. Fortunately, following *Citrullus colocynthis* intake, in their study, the damaged sinusoids, the haemorrhage and the inflammatory cell infiltration subsided indicating a beneficial role of such a remedy. These findings are supported by those claimed by Alarcon-Aguilar *et al.* (2000). Despite the obvious antidiabetic effect of *Citrullus colocynthis* pulp extract, its use in normal rats, in the current work, caused hepatocytic poration and few hapatocytes had condensed or fragmented nuclei indicating minimal hepatotoxicity.

This herbal therapy of Gymnema sylvestre (Asclepiadaceae) appears to bring about blood glucose homeostasis through increased serum insulin levels provided by repair/regeneration of the endocrine pancreas. Also, the effectiveness of the water extract from the leaves of Gymnema sylvestre, in controlling hyperglycemia was investigated in type II diabetic patients as conventional per os anti-hyperglycemic agents (Baskaran et al., 1990). The obtained data suggested that the β cells may be regenerated / repaired in type II diabetic patients on the extract's supplementation. This was supported by the raised insulin levels in the serum of patients after the supplementation. Furthermore, the antihyperglycemic action of a crude saponin fraction and five triterpene glycosides (gymnemic acids I-IV and gymnemasaponin V) derived from the methanol extract of leaves of Gymnema sylvestre was investigated in STZ-diabetic mice (Sugihara et al., 2000).

The results indicated that insulin-releasing action of gymnemic acid IV, administered i.p., may contribute to the antihyperglycemic effect by the leaves of Gymnema sylvestre.

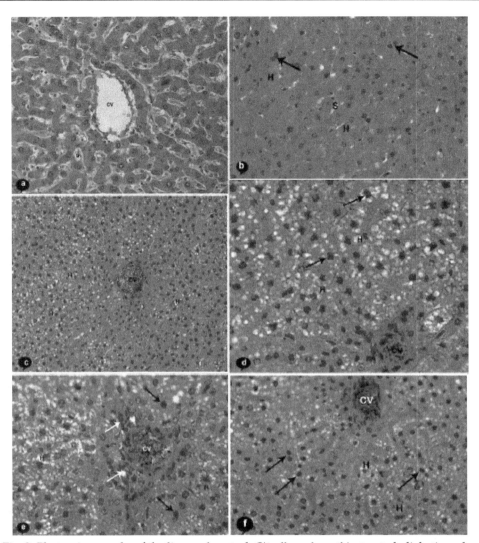

Fig. 3. Photomicrographs of the livers of control, *Citrullus colocynthis*- treated, diabetic and *Citrullus colocynthis*treated diabetic rats, H&E stain; (a) Liver of control rats, ×40; (b) Liver of *Citrullus colocynthis* treated rats, showing few hapatocytes had condensed or fragmented nuclei ×40; (c) Liver of diabetic rats, showing disorganized hepatic cords, reduced sinusoids and many hepatocytes having cytoplasmic vacuolization, ×20; (d) Higher magnification of (c), showing most hepatocytes with cytoplasmic vacuolar degeneration and pyknotic nuclei (arrows), ×40; (e) Liver diabetic rats, showing a central inflammatory cell infiltration (white arrows) and hepatocytes with cytoplasmic vacuolar degeneration and pyknotic nuclei (black arrows), ×40; (f) Liver of diabetic rats treated with *Citrullus colocynthis*, showing recovered hepatocytes with less cytoplasmic vacuolization compared with diabetic animals. Few hepatocytes with cytoplasmic vacuolization and pyknotic nuclei are still seen (arrows), ×40. According to the study occurred by Mohammad Khalil, et al. (2010).

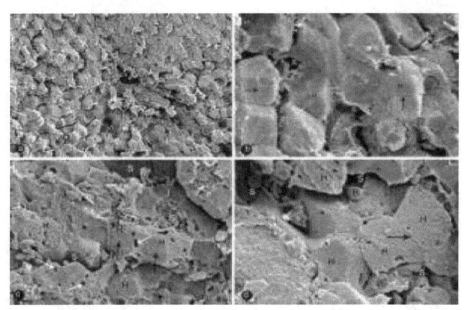

Fig. 4. Scanning electron micrographs of livers of control and *Citrullus colocynthis*-treated rats. (a) SEM of the liverof control rats showing normal hepatic cord around a Central Vein (CV), ×450; (b) Larger magnification of (a) showing normal Sinusoids (S) and normal Hepatocytes (H) with normal intercellular boundaries (arrows), ×1500 (c); SEM of the liver of *Citrullus colocynthis*-treated rats, showing rows of Hepatocytes (H) containing few pores (white arrows). Note the blood Sinusoids (S) and the intercellular boundaries (black arrows), ×1100; (d) Higher magnification of (c), showing blood Sinusoids (S), Hepatocytes (H), intercellular spaces (black arrows) and erythrocytes (R), ×1500. According to the study occurred by Mohammad Khalil, et al. (2010).

Gymnemic acid IV may be an anti-obese and antihyperglycemic prodrug. The inhibitory activity of each triterpene glycoside on the glucose uptake in rat small intestine fragments was examined, in order to determine its impact on the increase of serum glucose level in glucose-loaded rats (Yoshikawa et al., 1997). It was found that Gymnemic acids II and III showed potent inhibitory activity on glucose uptake. Gymnemoside b and gymnemic acids III, V, VII were found to exhibit a little inhibitory activity against glucose uptake, and the principal constituents, gymnemic acid I and gymnemasaponin V, lacked this activity. It is noteworthy that, although Gymnema saponin constituents such as Gymnemic acids II and III show no effect on serum glucose levels in oral-loaded rats, they exhibit potent inhibitory activity on the glucose uptake and further studies need to be contacted.

The hypoglycemic activity of a decoction from *Juniperus communis* (juniper berries) both in normal glycemic and in STZ-diabetic rats was studied (Sanchez et al., 1994). Juniper decoction decreased glycemic levels in normal glycemic rats through an increase of peripheral glucose utilization or a potentiation of glucose induced insulin secretion.

The per os administration of the decoction to STZ diabetic rats resulted in a significant reduction both in blood glucose levels and in the mortality index, as well as the prevention

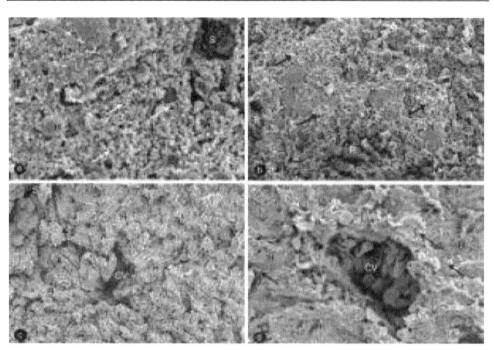

Fig. 5. Scanning electron micrographs of livers of diabetic and *Citrullus colocynthis*-treated diabetic rats. (a) SEM of the liver of diabetic rats showing degenerating Hepatocytes (H) having a plenty of lipid droplets (white arrows). Damaged blood Sinusoids (S) are also seen as well as some erythrocytes (R) are spread inside hepatocytes, ×450; (b) SEM of the liver of diabetic rats showing hemorrhage of erythrocytes (R) between Hepatocytes (H). Lipid degeneration (black arrows) of hepatocytes is also seen, ×450; (c) SEM of the liver of *Citrullus colocynthis*-treated diabetic rats, showing intact Hepatocytes (H) and blood Sinusoids (S). Only fewhepatocytes around the central vein have some vacuoles, ×450; (d) SEM of the liver of *Citrullus colocynthis* treated diabetic rats (300 mg) with higher magnification, showing intact Hepatocytes (H) and a Central Vein (CV) containing a lot of erythrocytes. A few hepatocytes around the central vein have a few lipid droplets (arrows), ×1100. According to the study occurred by Mohammad Khalil, et al. (2010).

of the loss of body weight. This effect seemed to be mediated by the peripheral action of juniper while in Olea europaea (Oleaceae) plant species, the hypoglycemic activity may result from two mechanisms: (a) potentiation of glucose induced insulin release, and (b) increased peripheral uptake of glucose.

The hypoglycemic effect of the rhizomes of *Polygala senega* (Polygalaceae) was proposed that after i.p. administration was without altering the insulin levels and with the need of the presence of insulin in order to act. In addition, one of the active components of this hypoglycemic effect was identified as a triterpenoid glycoside, senegin-II (as a mixture of isomers). *Solanum lycocarpum (Solanaceae)* has been widely employed for diabetes management, obesity and to decrease cholesterol levels. Some of the polysaccharides slowed gastric emptying and act on the endocrinous system affecting the liberation of gastrointestinal hormones, lowering blood glucose levels. The hypocholesterolemic activity could be due to the increased fecal bile acid excretion as well as to the action of the short-chain fatty acids, coming from fermentation, on the synthesis of deltaaminolevulinate and by the increase of the cholesterol 7-alpha-hydroxylase and 3-hydroxy-3-methylglutaryl CoA reductase synthesis (DalAgnol & Lino von Poser 2000).

4. Conclusions and future directions

Scientists from divergent fields are investigating plants anew with an eye to their antimicrobial usefulness. A sense of urgency accompanies the search as the pace of species extinction continues. Laboratories of the world have found literally thousands of phytochemicals which have inhibitory effects on all types of microorganisms in vitro. More of these compounds should be subjected to animal and human studies to determine their effectiveness in whole-organism systems, including in particular toxicity studies as well as an examination of their effects on beneficial normal microbiota. It would be advantageous to standardize methods of extraction and in vitro testing so that the search could be more systematic and interpretation of results would be facilitated. Also, alternative mechanisms of infection prevention and treatment should be included in initial activity screenings. Disruption of adhesion is one example of an anti-infection activity not commonly screened for currently. Attention to these issues could usher in a badly needed new era of chemotherapeutic treatment of infection by using plant-derived principles.

5. References

Aburjai, T., M. Hudaib, R. Tayyem, M. Yousef and M. Qishawi, 2007. Ethno-pharmacological survey of medicinal herbs in Jordan, the Ajloun heights egion. J. Ethnopharmacol., 110: 294-304. PMID: 7097250

Alarcon-Aguilar, F.J., M. Jimenez-Estrad, R. Reyes- Chilpa and R. Roman-Ramos, 2000. Hypoglycemic effect of extracts and fractions from Psacalium decompositum in healthy and alloxan-diabetic mice. J. Ethnopharmacol., 72:

Arise, R.O., S.O. Malomo, J.O. Adebayo and A. Igunnu, 2009. Effects of aqueous extract of eucalyptus globules on lipid peroxidation and selected enzymes of rat liver. J. Med. Plants Res., 3: 77-81

Brixova, E., 1981. Experimental and clinical liver steatosis. Folia Fac. Med. Univ. Comenian Bratisl., 19: 9-90.

Ban, C.R. and S.M. Twigg, 2008. Fibrosis in diabetes complications: Pathogenic mechanisms and circulating and urinary markers. Vasc. Health Risk Manage., 4: 575-596. PMID: 18827908

Bolkent S, Yanardag R, Tabakoglu-Oguz A, Ozsoy-Sacan A. "Effects of chard (Beta bulgaris L. var. cicla) extract on pancreatic B cells in diabetic rats: a morphological and biochemical study". *J. Ethnopharmacol.* 2000; 73: 251-259.

Baskaran K, Kizar Ahamath B, Radha Shanmugasundaram K, ShanmugasundaramER. "Antidiabetic effects of a leaf extract from Gymnema sylvestre in non-insulin-dependent diabetes mellitus patients". *J. Ethnopharmacol.* 1990; 30(3): 295-300.

Bujanda, L., E. Hijono, M. Larzabal, M. Beraza and P. Aldazabal *et al.*, 2008. Resveratrol inhibits non-alcoholic fatty liver disease in rats. BMC Gastroenterol., 8: 40-40. DOI: 10.1186/1471-230X-8-40

Chang SL, Chang CL, Chiang YM, Hsieh RH, Tzeng CR, Wu TK, Sytwu HK, Shyur LF, Yang WC. "Polyacetylenic compounds and butanol fraction from Bidens pilosa can modulate the differentiation of helper T cells and prevent autoimmune diabetes in non-obese diabetic mice". *Planta Med.* 2004; 70(11): 1045-1051.

Chen TH, Chen SC, Chan P, Chu YL, Yang HY, Cheng JT. " Mechanism of the hypoglycemic effect of stevioside, a glycoside of Stevia rebaudiana". *Planta Med.* 2005; 71(2): 108-113. Dall'Agnol R, Lino von Poser G. "The use of complex polysaccharides in the management of metabolic diseases: the case of Solanum lycocarpum fruits". *J. Ethnopharmacol.* 2000; 71(1-2): 337-341.

Fagot-Campagna A. & Narayan K. "Type 2 diabetes in children". *Br. Med. J.* 2001: 322: 377-387.

Fugh-Beerman A. "Herb-drug interactions". *Lancet* 2000; 355: 134-138.

Fujisawa T, Ikegami H, Kawaguchi Y, Nojima K, Kawabata Y, Ono M, Nishino M, Noso S. Taniguchi H, Horiki M, Itoi-Babaya M, Babaya N, Inoue K, Ogihata T. "Common genetic basis between type I and type II diabetes mellitus indicated by interview-based assessment of family history ". *Diabetes Res. Clin. Pract.* 2004; 66S:S91-S95

Ghosh, S. and S.A. Surawanshi, 2001. Effect of *Vinca rosea* extracts in treatment of alloxan diabetes in male albino rats. Indian J. Exp. Biol., 39: 748759.PMID:12018575

Grover JK, Vats V, Yadav SS. "Pterocarpus marsupium extract (Vijayasar) prevented the alteration in metabolic patterns induced in the normal rat by feeding an adequate diet containing fructose as sole carbohydrate". *Diabetes Obes. Metab.* 2005; 7(4): 414-417.

Huxtable RJ. "The harmfull potential of herbal and other plant products". Drug Safety 1990; 5(Suppl. 1): 126-136.

Hsu FL, Lai CW, Cheng JT. "Antihyperglycemic effects of paeoniflorin and 8-debenzoylpaeoniflorin, glycosides from the root of Paeonia lactiflora". *Planta Med.* 1997; 63(4): 323-325.

Hayden, M.R., J.R. Sowers and S.C. Tyagi, 2005. The central role of vascular extracellular matrix and basement membrane remodeling in metabolic syndrome in type and 2 diabetes: The matrix preloaded. Cardiovasc. Diabetol., 4: 9-9. DOI: 10.1186/1475-2840-4-9

Khan, Z.A. and S. Chakarabarti, 2003. Endothelins in chronic diabetic complications. Can. J. Physiol. Pharmol., 81: 622-634. DOI: 10.1139/Y03-053

Keith, K.G., V. Fonseca, M.H. Tan and A. Dalpiaz, 2004. Narrative review: Hepatobiliary disease in type 2 diabetes mellitus. Ann. Intern. Med., 141: 946-956. PMID: 15611492

Koyama Y, Abe K, Sano Y, Ishizaki Y, Njelekela M, Shoji Y, Hara Y, Isemura M. "Effects of green tea on gene expression of hepatic gluconeogenic enzymes in vivo". *Planta Med.* 2004; 70(11): 1100-1102.

Karageuzyan KG, Vartanyan GS, Agadjanov MI, Panossian AG, Hoult JR. "Restoration of the disordered glucose-fatty acid cycle in alloxan-diabetic rats by trihydroxyoctadecadienoic acids from Bryonia alba, a native Armenian medicinal plant". *Planta Med.* 1998; 64(5): 417-422.

Ji Su Kim, Jung Bong Ju , Chang Won Choi and Sei Chang Kim (2006): Hypoglycemic and Antihyperlipidemic Effect of Four Korean Medicinal Plants in Alloxan Induced Diabetic Rats American Journal of Biochemistry and Biotechnology 2 (4): 154-160, 2006 ISSN 1553-3468

Luo J, Chuang T, Cheung J, Quan J, Tsai J, Sullivan C, Hector RF, Reed MJ, Meszaros K, King SR, Carlson TJ, Reaven GM. "Masoprocol (nordihydroguaiaretic acid): a new antihyperglycemic agent isolated from the creosote bush (Larrea tridentata)". *Eur. J. Pharmacol.* 1998; 346(1): 77-79.

Luo J, Fort DM, Carlson TJ, Noamesi BK, nii-Amon-Kotei D, King SR, Tsai J, Quan J, Hobensack C, Lapresca P, Waldeck N, Mendez CD, Jolad SD, Bierer DE, Reaven GM. " Cryptolepis sanguinolenta: an ethnobotanical approach to drug discovery and the isolation of a potentially useful new antihyperglycemic agent". *Diabet. Med.* 1998; 15(5): 367-374.

Lyra, R., M. Oliveira, D. Lins and N. Cavalcanti, 2006. Prevention of type 2 diabetes mellitus. Arq. Bras. Endocrinol. Metabo,. 50: 239-249.

MacFarlance IA, Bliss M, Jackson JG, Williams G. "The history of diabetes mellitus. In textbook of diabetes". (eds Pichup, j. Williams G.) Blackwell, London 1997, 2nd edn, vol. I, pp 1-21.

Moller, D.E., 2001. New drug targets for type 2 diabetes and the metabolic syndrome. Nature, 414: 821-827. PMID: 11742415

Mohammad Khalil, Gamal Mohamed, Mohammad Dallak, Fahaid Al-Hashem, Hussein Sakr, Refaat A. Eid, Mohamed A. Adly, Mahmoud Al-Khateeb, Saleh Banihani, Zuhair Hassan and Nabil Bashir (2010):The Effect of *Citrullus colocynthis* Pulp Extract on the Liver of Diabetic Rats a Light and Scanning Electron Microscopic Study. American Journal of Biochemistry and Biotechnology 6 (3): 155-163, 2010 ISSN 1553-3468

Naik SR, Barbosa Filho JM, Dhuley JN, Deshmukh V. "Probable mechanism of hypoglycemic activity of bassic acid, a natural product isolated from Bumelia sartorum". *J. Ethnopharmacol.* 1991; 33(1-2): 37-44.

Pushparaj PN, Tan BK, Tan CH. "The mechanism of hypoglycemic action of the semi-purified fractions of Averrhoa bilimbi in streptozotocin-diabetic rats". *Life Sci.* 2001; 70(5): 535-547.

Rajasekaran S, Sivagnanam K, Ravi K, Subramanian S. "Hypoglycemic Effect of Aloe vera Gel on Streptozotocin-Induced Diabetes in Experimental Rats". *J. Med. Food.* 2004; 7(1): 61-66.

Rollo J. "An account of two cases of the diabetes mellitus, with remarks as they arose during the progress of the cure". C. Dilly, London, 1797.

Sanchez de Medina F, Gamez MJ, Jimenez I, Jimenez J, Osuna JI, Zarzuelo A."Hypoglycemic activity of juniper berries". Planta Med. 1994; 60(3): 197-200.

Sebbagh, N., C. Cruciani-Guglielmacci, F. Cruciani, Guglielmacci, M.F. Berthault and C. Rouch et al., 2009. Comparative effects of Citrullus colocynthis, sunflower and olive oil-enriched diet on streptozotocin-induced diabetes in rats. Diabetes Metab., 35: 178-184. PMID: 19264524

Seifalian, A.M., C. Piasecki, A. Agawel and B.R. Davidson, 1999. The effect of graded steatosis on flow in the hepatic on flow in the hepatic parenchymal microcirculation. Transplantation, 68: 780-784. PMID: 10515377

Sugihara Y, Nojima H, Matsuda H, Murakami T, Yoshikawa M, Kimura I."Antihyperglycemic effects of gymnemic acid IV, a compound derived fromGymnema sylvestre leaves in streptozotocin-diabetic mice". J. Asian Nat. Prod. Res. 2000; 2(4): 321-327.

Ugochukwu NH, Babady NE. "Antihyperglycemic effect of aqueous and ethanolic extracts of Gongronema latifolium leaves on glucose and glycogen metabolism in livers of normal and STZ-diabetic rats". Life Sci. 2003; 73(15): 1925-1938.

Ugochukwu NH, Babady NE. "Antioxidant effects of Gongronema latifolium in hepatocytes of rat models of non-insulin dependent diabetes mellitus". Fitoterapia 2002; 73: 612-618.

Ugochukwu NH, Cobourne MK. "Modification of renal oxidative stress and lipid peroxidation in STZ-diabetic rats treated with extracts from Gongronema latifolium leaves". Clinica Chimica Acta 2003; 336: 73-81

Venkateswaran S, Pari L. "Effect of Coccinia indica leaves on antioxidant status in STZ-diabetic rats". J. Ethnopharmacol. 2003; 84: 163-168.

Venkateswaran S, Pari L. "Effect of Coccinia indica on Blood Glucose, Insulin and Key Hepatic Enzymes in Experimental Diabetes". Int. J. Pharmacogn. 2002; 40(3): 165-170.

Waltner-Law, M.E., X.L. Wang, B.K. Law, R.K. Hall and M. Nawano et al., 2002. Epigallocatechin gallate, a constituent of green tea, represses hepatic glucose production. J. Biol. Chem., 277: 34933-34940. PMID: 12118006

WHO (1978a). Alma Ata Declaration. Primary Health Care. Health for all series No.1

World Health Organization, 1980. WHO expert committee on diabetes mellitus.

Witters L. "The blooming of the French lilac". J. Clin. Invest. 2001; 108: 1105-1107.

Yoshikawa M, Murakami T, Kadoya M,. "Medicinal foodstuffs. III. Sugar beet. (1): Hypoglycemic oleanolic acid oligoglycosides,betavulgarosides I, II, III, and IV, from the root of Beta vulgaris L. (Chenopodiaceae)". Chem. Pharm. Bull. (Tokyo) 1996; 44: 1212-1217.

Yoshikawa M, Murakami T, Kadoya M, Li Y, Murakami N, Yamahara J, Matsuda H. "Medicinal foodstuffs. IX. The inhibitors of glucose absorption from the leaves of Gymnema sylvestre R. BR. (Asclepiadaceae): structures of gymnemosides a and b". Chem. Pharm. Bull. (Tokyo). 1997; 45(10): 1671-1676.

Yoshikawa M, Shimada H, Nishida N, Li Y, Toguchida I, Yamahara J, Matsuda H. "Antidiabetic principles of natural medicines. ALR2 and alpha-glucosidase inhibitors from Brazilian natural medicine, the leaves of Myrcia multiflora DC. (Myrtaceae): structures of myrciacitrins I and II and myrciaphenones A and B". *Chem. Pharm. Bull. (Tokyo).* 1998; 46(1): 113-119.

7

Red Palm Oil and Its Antioxidant Potential in Reducing Oxidative Stress in HIV/AIDS and TB Patients

O. O. Oguntibeju, A. J. Esterhuyse and E. J. Truter
Oxidative Stress Research Centre, Department of Biomedical Sciences,
Faculty of Health & Wellness Sciences, Cape Peninsula University of Technology,
South Africa

1. Introduction

1.1 HIV and TB

Scientific evidence has shown that HIV infection is caused by a retrovirus, the Human Immunodeficiency Virus (HIV) which is a ribonucleic acid (RNA) virus so designated because of its genome that encodes an unusual enzyme, reverse transcriptase (RT) that enables the virus to make copies of its own genome as DNA in its host's cells (human T4 helper lymphocytes) (Oguntibeju et al., 2008).

The drastic increase in the number of people infected with HIV is not peculiar to a particular racial group, country or community despite multidimensional efforts which have been made to combat this scourge (Weiss, 1996; Oguntibeju et al., 2007a). It is reported that the virus selectively attacks and depletes T-lymphocyte bearing CD4+ cells (T-helper cells) causing a predisposition to opportunistic infections and malignancies (Weiss, 1996) and ultimately resulting in Acquired Immunodeficiency Syndrome (AIDS).

The cellular receptors to HIV are cells that express the CD4+ T cell receptor (CD4+ T-cells or T4-cells) as well as other white blood cells including monocytes and macrophages. Glial cells in the central nervous system, chromaffin cells in the intestine and Langerhans cells in mucous membranes and skin that express CD4+ T cell receptors can also be infected (Paxon et al., 1996). The possibility that there are other cellular targets apart from CD4+T-cells is proved by the likelihood of neurons that can be infected. This creates the possibility of the presence of co-receptors in addition to CD4+ T cells to mediate fusion between HIV and its target cells (Grossman and Heberman, 1997).

Recognition of the CD4+ T-cells by HIV-1 envelope glycoprotein (gp120) to which the virus binds and enters host cells to initiate rapid replication cycles (Oguntibeju et al., 2007b) depicts significant cytopathic consequences of HIV infection of CD4+ T-cells (Bartlett, 1998) and is an important factor in the initiation of HIV infection. The shed virions which are immunogenic, stimulate B cells to produce humoral antibodies and plasma cells through lymphoid hyperplasia that ultimately results in decreased number of infected cells as the CD4+ T-cells migrate through the germinal cells. The depletion in the number of CD4+ T-cells exceeds the formation of new cells and may maintain this phase for many years resulting in general disorganization of the lymphoid nodes, loss of lymphoid function and integrity.

2. Physiological and biochemical mechanisms of the role of oxidative stress in HIV/AIDS & TB disease complications

After initial infection of the human host, the pace of immunodeficiency development, susceptibility to infection and malignancies become manifest and are generally associated with the rate of CD4[+] decline (Enger et al., 1996). The rate of CD4[+] decline varies considerably from person to person and is not constant throughout all the stages of HIV infection. Though the virological and immunological process that take place during the period of rapid fall in the number of CD4[+] T-cells are poorly understood, Koot et al. (1996) reported that acceleration of the decline of CD4[+] T-cells heralds the progression of the disease that is associated with the increasing rate of HIV-1 replication *in vivo* and declining cell-mediated immune response. Studies have shown that the host immunological alterations due to HIV infection result in progressive development of opportunistic infections and malignancy and is chiefly mediated /induced by deregulation of a cytokine profile production of ROS which also plays a role in the viral replication. *In vitro* studies have shown activation of viral replication by induction of TNFαβ (Allard et al., 1998).

Das and co-workers (1990) stated that excessive production of reactive oxygen species (ROS) such as superoxide anions, OH- radicals and H_2O_2, may be related to increased activation of PMN leucocytes during infection. This is influenced by the pro-oxidant effect of TNFα produced by the activated macrophages during the course of HIV infection and secretion of pro-inflammatory cytokines IL-1, IL-6, and IL-8 (Kiedzierska and Crowe, 2001). Gil et al. (2003) further established the presence of substantial oxidative stress in HIV infection which they attributed to the role of viral proteins that increases ROS intracellularly, therefore increasing the apoptotic index and depleting the CD4[+] T-lymphocyte population. The ROS thus produced can attack the double bonds in polyunsaturated fatty acids, inducing lipid peroxidation which may result in more oxidative cellular damage to the membrane lipids, proteins or DNA. Chronic oxidative stress experienced by patients infected by HIV leads to a condition in which there is increased consumption of antioxidants (such as Vitamin C, and E , selenium, and carotenoids) as well as micronutrients/ trace elements (Januga et al., 2002). Stephen (2006), therefore, concluded that persistent chronic inflammation such as found in HIV infection places a long-term strain on antioxidant defenses, impair immune functions, increases the severity of the disease as well as increases the antioxidant requirement by the infected individual.

Progression of HIV to AIDS in developed countries after initial infection is about 10-12 years for adults in the absence of antiviral therapy. However, some individuals manifest full blown AIDS within 5 years of infection whereas others survive long term (>10 years) asymptomatic HIV-1 infection without a significant decline in CD4[+] T-cell count. Such delay in the progression of HIV to AIDS may be attributed to either infection with genetically defective HIV-1 variants or effective host antiviral immune response where the individual has active cytotoxic T-cell responses against HIV-1 infected cells (Haase, 1999).

Ever since Robert Koch made the landmark discovery that tuberculosis is caused by the infectious agent *Mycobacterium tuberculosis* (Koch, 1882), it has remained a major global health threat. Although in developed countries the rates of infection has fallen in the past century, the number is now again increasing which results in over 2000 deaths in developed countries annually due to changes in social structures in cities, the HIV epidemic, and failure to improve treatment programs (Frieden et al., 1995). The increased death rate recorded as a result of poverty, poor living conditions and inadequate medical care in developing/Third World countries is further compounded by the emergence of multi-drug resistance where antibiotics are either of inferior quality, or are not used for a sufficient period of time to control the disease (O'Brien, 2001).

The recent increase in reported pulmonary tuberculosis (PTB) cases globally can be attributed to the increased susceptibility to opportunistic infections in HIV-infected persons. The highest prevalence of cases is reported to be in Asia (China, India, Indonesia, Bangladesh and Pakistan) and Africa with over 90% of global TB infections and deaths annually. TB cases occur predominantly in the economically productive 15-49 year age group (Dye *et al.*, 1999). Like HIV infection, TB also has a long latency period with symptomatic presentation occurring from 3 months to decades after establishment of the infection (Jagirdar and Zagzag, 1996).

TB is caused by an obligate pathogen that does not replicate outside its host environment (Mathema *et al.*, 2006) and is spread by aerosolization of droplets bearing *M. tuberculosis* particles released from the lung or larynx during coughing, sneezing, or talking in poorly ventilated areas. The particles of 1-5 μm in diameter, are inhaled and phagocytosed by resident alveolar macrophages. A vigorous immune response involving cytokines and a large number of chemokines ensues (Roach *et al.*, 2002). Protective immunity is characterized by granuloma formation that consists of primarily activated *M. tuberculosis* infected macrophages and T-cells. Medlar (1955), noted tissue necrosis and cavitations in over 10% of presumed immuno-competent patients and postulated that this was due to non-containment of continual bacterial replication (doubling time of 25-32 hours) that resulted in disease symptoms and its associated pathology. This response presumably initially limits infection to the primary site of invasion (the lung parenchyma and local draining lymph nodes known as the Ghon complex) in the majority of immuno-competent individuals (Bloom and Murray, 1992,).

Increased reactive oxygen species (ROS) has been reported in patients with TB. Excessive endogenously produced ROS in activated phagocytes of TB patients that escape to its surroundings can damage tissue or cellular DNA as well as impair immune function (Madebo *et al.*, 2003). It has been shown that the bactericidal potency of the myeloperoxidase-H_2O_2-halide system of neutrophilic granules demonstrates the bactericidal activities of the phagocytes that invariably produce increased ROS and reactive nitrogen intermediates (RNI) during phagocytic respiratory burst. Lower antioxidant potential as shown by a significant reduction of enzymatic antioxidants (superoxide dismutase, catalase) and non-enzymatic antioxidants (glutathione) as well as high malondialdehyde (MDA) concentrations suggest increases in the generation of ROS due to lipid peroxidation (Reddy *et al.*, 2004).

Di Massio and co-workers (1991), reported significantly reduced vitamin C and α-tocopherol levels in TB patients. These are integral components of antioxidants, which, when present in sufficient quantity, may act synergistically to protect cells from oxidative stress induced damage in TB patients. Several factors such as inadequate nutrients, malnutrition, nutrient malabsorption, low food intake, inadequate nutrient release from the liver, acute infections including other than HIV, may be the cause of low or impaired antioxidant capacity in TB patients (Das *et al.*, 1990).

Presentation of disease is variable as regards the pathology as well as infections in a variety of tissues such as the meninges, lymph nodes, and tissue of the spine, where response to antibiotic medication/treatment to clear the bacilli from tissues, partial reversal of the granulomatous process and subsequent clearance from the sputum may be found in clinical cases (Jargirdar and Zagzag, 1996). The progression and nature of disease may be affected by factors such as conditions that negatively impact on the host immune system (for example, poorly controlled diabetes mellitus, renal failure, chemotherapy, malnutrition or intrinsic host susceptibility (Madebo *et al.*, 2003). Host susceptibility has been known to affect endogenous re-activation and exogenous re-infection by the bacilli.

3. Reactive oxygen species and reactive nitrogen species and their effects on biological macromolecules and organs

Reactive oxygen species / reactive nitrogen species (ROS/ RNS) are constantly being formed in living organisms (Ceconi *et al.*, 2003). In the course of oxygen metabolism, 1- 5% of all inhaled oxygen becomes ROS (Berk, 2007). Endogenously, ROS are produced from various sources such as mitochondria, activated macrophages and leucocytes, oxidase enzyme (NADPH), cyclo-oxygenase and lipoxygenase (Zalba *et al.*, 2006). Reactive oxygen species have oxidation ability and are classified either as free radicals (superoxide anion O_2^{-}, hydroxyl radical $OH^{\bullet}$, nitric oxide NO) or as non-free radicals (hydrogen peroxide H_2O_2, peroxynitrite $ONOO^{-}$) (Higashi *et al.*, 2006). Previous studies have shown the involvement of ROS in physiological and pathophysiological conditions (Fortuño *et al.*, 2005; Berk, 2007; Heistad *et al.*, 2009). At low concentrations, ROS are involved in normal cell signaling pathways (smooth muscle and endothelial cell growth, apoptosis and survival) and in the remodeling of vessel walls (Fortuño *et al.*, 2005; Heistad *et al.*, 2009). At high concentrations, ROS are identified as harmful compounds and constitute an important risk factor for the development of many diseases such as cardiovascular diseases (Maxwell & Lip, 1997; Heistad *et al.*, 2009).

4. Oxidative stress

Oxidative stress occurs when there is a dysfunction in the overall balance between the production of reactive oxygen and nitrogen species and the antioxidant defense mechanisms (Ceconi *et al.*, 2003; Berk, 2007; Barbosa *et al.*, 2008).

Oxidative stress is believed to play a critical role in the complications and pathophysiology of HIV/AIDS, TB and cardiovascular diseases (Heistad *et al.*, 2009). In the context of oxidative stress in HIV/AIDS and TB, the major vascular ROS is the superoxide anion (O_2^{-}) which is predominantly generated by the NADPH oxidase enzyme (Fortuño *et al.*, 2005). Superoxide is normally dismutased to hydrogen peroxide (H_2O_2) by a family of superoxide dismutase (intracellular Cu/Zn SOD, MnSOD or extracellular Cu/Zn SOD) (Hamilton *et al.*, 2004). Hydrogen peroxide is converted into oxygen and water by catalase enzymes or by glutathione peroxidase (GPx) in the presence of reduced glutathione (Hamilton *et al.*, 2004; Zalba *et al.*, 2006). In the pathophysiological process of oxidative stress, excess superoxide has many effects, superoxide combines with NO to form peroxynitrite. Peroxynitrite is a highly toxic oxidant which causes damage to cells of the vascular wall through oxidation of lipids (lipid peroxidation), proteins (protein nitrosilation) and nucleic acids with superoxide. This causes vascular dysfunction by removing the protective effects of NO (Heistad *et al.*, 2009), initiates the development of vascular inflammatory state (Hamilton *et al.*, 2004), facilitates the oxidation of LDL, causing development of artherosclerotic lesions (Zalba *et al.*, 2006) and triggers apoptotic cell death (Ceconi *et al.*, 2003).

Accumulating evidence has suggested that oxidative stress, mainly through lipid peroxidation, represents an important risk factor in the development of cardiovascular diseases and complications in HIV/AIDS and TB (Waterfall et al., 1997; Ceconi et al., 2003). In fact, lipid peroxidation leads to membrane disruption and release of highly reactive free radicals (such as MDA) which can severely alter the cellular function (Ceconi et al., 2003) (Table 1).

S/N	Oxidants	Reactions
1.	Production of superoxide	O_2 + electron $\longrightarrow$ O_2
2.	NADPH – oxidase	$2O_2$ + NADPH $\longrightarrow$ $2O_2^-$ + NADP + H^+
3.	Superoxide dismutase	O_2^- + O_2^- + $2H^+$ $\longrightarrow$ H_2O_2 + O_2
4.	Calalase	H_2O_2 $\longrightarrow$ $2H_2O$ + O_2
5.	Myeloperoxidase	H_2O_2 + x^- + H^+ $\longrightarrow$ HOX + H_2O
6.	Glutathione peroxidase (Se-dependant)	2GSH + R-O-OH $\longrightarrow$ GSSG + H2O + ROH
7.	Fenton reaction	Fe^{2+} + H_2O_2 $\longrightarrow$ Fe^{3+} OH + OH^-
8.	Iron-catalyzed Haber Weiss reaction	O_2^- + H_2O_2 $\longrightarrow$ O_2 + OH + OH^-
9.	Glucose-6-phosphate dehydrogenase	G-6-P + NADP $\longrightarrow$ 6-Phosphogluconate + NADPH + H^+
10.	Glutathione reductase	G-S-S-G+NADPH + H^+ $\longrightarrow$ 2GSH + NADH

Source: Murray, 2000

Table 1. Reactions in relation to oxidative stress in blood cells and various tissues.

5. Activation of oxygen

Oxygen is essential for energy metabolism and respiration but is has been implicated in many disease and degenerative conditions (Ceconi et al., 2003). Activation of oxygen may occur by two different mechanisms: absorption of sufficient energy to reverse the spin on one of the unpaired electrons and monovalent reduction. Non-activated oxygen is a bi-radical. It can be activated by either reversing the spin on one of the unpaired electrons to form the singlet state or by reduction. In the monovalent reduction of oxygen, superoxide (O_2^-), hydrogen peroxide (H_2O_2), hydroxyl radical (OH) and finally, water (H_2O) is formed. Superoxide forms the hydroxyl radical (OOH) which is the protonated form of the superoxide anion radical (Gebick and Bielski, 1981; Ceconi et al., 2003).

Numerous enzymes (peroxidases) use hydrogen peroxide as a substrate in oxidation reactions involving the synthesis of complex organic molecules. Haber and Weiss (1994) identified the hydroxyl radical as the oxidizing species in the reaction between H_2O_2 and ferrous salts.

$$Fe^{2+} + H_2O_2 \longrightarrow Fe^{3+} + OH + OH^-$$

Most oxygen is consumed by the cytochrome oxidase enzyme in the mitochondrial electron transport system. Isolated mitochondria produce H_2O_2 and O^{2-} in the presence of NADH (Loschen *et al.*, 1974). The various Fe-S-proteins and NADH dehydrogenase have also been implicated as possible sites of superoxide and hydrogen peroxide formation (Waterfall *et al.*, 1997). Various oxidative processes including oxidation hydroxylations, dealkylations, deaminations, dehalogenation and desaturation occur in the smooth endoplasmic reticulum. Mixed function oxygenases that contain a heme moiety add an oxygen atom into an organic substrate using NADPH as the electron donor. The generalized reaction catalyzed by cytochrome P_{450} is:

$$RH + NADPH + H^+ + O_2 \longrightarrow ROH + NADP^+ + H_3O$$

Superoxide is produced by microsomal NADPH dependent electron transport involving cytochrome P_{450} (Valko *et al.*, 2007). One possible site at which this may occur is shown in Figure 1:

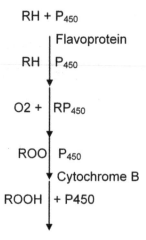

Fig. 1. Schematic presentation of the cytochrome P450 electron transport (Valko *et al.*, 2007).

In the peroxisomes and glyoxysomes, compartmentalized enzymes involved in the B-oxidation of fatty acids and glyoxylic acid cycle includes glycolate oxidase, catalase and various peroxidases. Glycolate oxidase produces H_2O_2 in a two-electron transfer from glycolate to oxygen (Lindqvist *et al.*, 1991). Xanthine oxidase, urate oxidase and NADH oxidase generate superoxide as a consequence of the oxidation of their substrates (Fridovich, 1970). The xanthine oxidase reaction is often used *in vitro* as a source of superoxide producing one mole of superoxide during the conversion of xanthine to uric acid (Fridovich, 1970). A superoxide-generating NADPH oxidase activity has been clearly identified in plasmalemma-enriched fractions (Valko *et al.*, 2007). These flavopoteins may produce superoxide by the redox cycling of certain quinones or nitrogenous compounds and NADPH oxidase reduces Fe^{3+} to Fe^{2+} converting it to a form that can be transported. Dysfunction of NADPH oxidase results in the formation of superoxide (Maxwell & Lip, 1997).

6. Mechanism of action of antioxidants in improving the immune status of AIDS and TB patients

Antioxidants are compounds that dispose, scavenge and suppress the formation of free radicals or oppose their actions. Free radicals, primarily the ROS, superoxide and hydroxyl radicals, which are highly reactive, with an unpaired electron in an atomic or molecular orbit, are generated under physiological conditions during aerobic metabolism (Semba & Tang, 1999; Champe et al., 2005). Because free radicals are potentially toxic, they normally are inactivated or scavenged by antioxidants before they can damage lipids, proteins or nucleic acids.

Superoxide dismutase (SOD), catalase and glutathione peroxidase (GSH-Px) are the primary antioxidant enzymes involved in the direct elimination of ROS whereas glutathione transferase, glucose-6-phosphate dehydrogenase (G6PD) and copper-binding ceruloplasmin are secondary antioxidant enzymes which assist in maintaining a steady concentration of glutathione and NADPH required for optimal functioning of the primary antioxidant enzymes (Kiremidjian-Schumacher et al., 1994; Champe et al., 2005). Antioxidant enzymes require micronutrients such as selenium, iron, copper, zinc and manganese as cofactors for optimal catalytic activity and to act as effective antioxidant defence mechanisms. If homeostasis between the rate of formation of free radicals and the rate of neutralization of free radicals is not maintained, oxidative damage, known as oxidative stress occurs which further damages the already compromised immune system and consequently enhances HIV and TB progression (Cunningham-Rundles, 2001; Cunningham-Rundles et al., 2005) and it has been reported that most of these antioxidants are derived from dietary sources (fruits and vegetables including red palm oil (Maxwell & Lip, 1997; Ebong et al., 1999; Edem, 2002; Van Rooyen et al., 2008).

7. Mechanism of action of red palm oil in improving the immune status of patients with AIDS and TB

Several studies have illustrated that red palm oil (RPO) is a rich cocktail of lipid-soluble antioxidants such as carotenoids (mostly α- and β-carotene, lycopenes), vitamin E (in the form of α- , β-, δ- tocotrienols and tocopherol) and ubiquinone (mostly coenzyme Q_{10}) (Ebong et al., 1999; Edem, 2002; Van Rooyen et al., 2008).

Feeding experiments using various animal models have highlighted that red palm oil is beneficial to health by reducing oxidative stress (Ebong et al., 1999). Many studies have demonstrated the protective effects of red palm oil in a cardiac ischaemia/perfusion model of oxidative stress (Esterhuyse et al., 2005; Bester et al., 2006; Engelbrecht et al., 2006) and modulation of the serum lipid profile in rats.

RPO is widely used as cooking oil in West and Central Africa and plays an essential role in meeting energy and essential fatty acid needs in many regions of the world. It contains many beneficial antioxidants and micronutrient compounds such as tocopherol, tocotrienol, lycopene, squalene, co-enzyme Q_{10}, physterol, glycolipids, phosphatides, calcium, phosphorus, iron, riboflavin, chlorophil, xanthophil, flavonoids, phospholipids, and carotenoid in addition to the equal proportion of saturated and unsaturated fatty acids such as oleic acid, linolenic acid, palmitic acid, linoleic acid, stearic acid and arachidic acid. It is known to be the richest source of carotenoids in terms of provitamin A equivalents i.e α and β carotenes (Sundram et al., 2003) with its wide range of protective properties against disease and aging as well as being modulators for cellular processes / functions where photo-oxidative processes predominate by acting as scavengers of oxygen and peroxyl radicals (Van Rooyen *et al.*, 2008). Sebinova and co-workers (1991) documented the increased protection derived from a combination of tocopherol and tocotrienol and further revealed that tocotrienol offers a more efficient protection than tocopherol as it is preferentially consumed by ROS. It has been shown that fresh RPO has no adverse effect on body weight and morphology of body tissues. It also lowers the level of serum lipids and inhibits tumour growth (Kritchevsky, 2000), enhances intestinal uptake of protein and the metabolism of sulphur-amino acids and promotes reproductive capacity (Ebong *et al.*, 1999). Calcium, phosphorous, iron, riboflavin, chlorophyll, xanthophylls, flavonoids and phospholipids and equal proportion of saturated and unsaturated fatty acids have also been identified as part of its constituents (Sundram *et al.*, 2003).

A number of human feeding studies reported that palm oil diets show a reduction of blood cholesterol values ranging from 7%-38% (Mattson and Grundy, 1985; Bonanome and Grundy, 1988). A comparative study in young Australian adults showed that the total blood cholesterol, triglycerides and HDL-cholesterol levels of those fed on palm oil (palm olein) and olive oil were lower than those fed on the usual Australian diet (Choudhury *et al.*, 1995). A double-blind cross-over study (Sundram, 1997) showed that a palm olein rich oil diet is identical to an oleic-acid rich diet.

A study on fifty-one Pakistani adults showed that those given palm oil rich diets performed better than those on sunflower oil. Palm oil was found to increase HDL-cholesterol and Apo A-1 levels (Farooq *et al*, 1996). A group in Beijing, China compared the effects of palm oil, soybean oil, peanut oil and lard (Zhang *et al*, 1997a; Zhang *et al*, 1997b) and reported that palm oil caused a significant increase in the HDL-cholesterol as well as a significant reduction in LDL-triglycerides.

Sundram and co-workers (1992) performed a dietary intervention study on a free-living Dutch population who normally consumes diets high in fats. Using a double blind cross-over study design consisting of two periods of six weeks of feeding, the normal fat intake of a group of 40 male volunteers was replaced by 70% palm oil. The palm oil diet did not raise serum total cholesterol and LDL-cholesterol, and caused a significant increase in the HDL-cholesterol as well as a significant reduction in LDL-triglycerides.

The effects of palm olein and of canola oil on plasma lipids were examined in double blind experiments in healthy Australian adults (Truswell *et al*, 1992). Palm oil performed better than canola oil in raising the HDL-cholesterol (Truswell *et al*, 1992). Other studies have demonstrated that RPO supplementation has beneficial or neutral effects on serum total cholesterol (Zhang *et al.*, 1997a).

A cross-over feeding study showed that the blood cholesterol, triglycerides, HDL-cholesterol and LDL-cholesterol levels of palm olein and olive oil diets were comparable (Ng *et al.*, 1992). A Malaysian study (Ng *et al.*, 1991) was conducted to compare the effects of a diet containing palm oil (olein), corn oil and coconut oil on serum cholesterol. Coconut oil was found to raise serum total cholesterol by more than 10% whereas both the corn and palm oil diets reduced the total cholesterol; the corn oil diet reduced the total cholesterol by 36% and palm oil by 19%. A similar cholesterol-lowering effect of palm oil was observed in 110 students in a study conducted in Malaysia (Marzuki *et al.*, 1991). The study compared the effect of palm oil with that of soybean oil. Volunteers fed on palm oil (olein) and soybean oil for five weeks, with a six-week wash-out period, showed comparable blood cholesterol levels. However, the blood triglycerides were increased by 28% in those on the soybean oil diet.

Thus, the impact of palm oil on serum lipids is more like that of a mono-unsaturated rather than saturated oil. There appears to be several explanations: (1) Palm oil is made up of 50% unsaturated fats and the saturated fatty acids present are palmitic (90%) and stearic (10%). Stearic acid as well as palmitic acid do not raise blood cholesterol levels in people whose blood cholesterol levels are in the normal range (Hayes, 1993; Hayes et al., 1995; Hayes et al., 1991, Khosla and Hayes, 1994; Khosla and Hayes, 1992). (2) The vitamin E, particularly the tocotrienol present in palm oil can suppress the synthesis of cholesterol in the liver (Qureshi et al., 1991a; Qureshi et al, 1991b; Qureshi et al, 1980; Mcltosh et al, 1991). (3) The position of the saturated and unsaturated fatty acid chains in a triglyceride backbone of the palm oil molecule determines whether the fat will elevate the cholesterol level in the blood (Kritchevsky, 1988; Kritchevsky, 1996). In palm oil, 87% of the unsaturated fatty acid chains are found in position 2 of the carbon atom of the triglyceride backbone molecule (Ong & Goh, 2002). This could explain why palm oil is not cholesterol-elevating. (4) It has an anti-clotting effect and prevents the formation of thrombi in the blood vessels. Hornstra (1988) in the Netherlands first demonstrated that palm oil has an anti-clotting effect, and is as anti-thrombotic as the highly unsaturated sunflower seed oil. A human study (Kooyenga et al., 1997) showed that tocotrienol (from palm oil) supplementation can reduce stenosis of patients with carotid atherosclerosis.

Holub et al, (1989) reported that the vitamin E in palm oil inhibits platelets from "sticking" to each other. Other supporting evidence showed that a palm oil diet increases the production of a hormone, prostacyclin that prevents blood-clotting or decreases the formation of a blood-clotting hormone, thromboxane (Sundram et al., 1990; Ng et al., 1992).

Kurfeld et al. (1990) in the United states, using a rabbit model, compared the effects of palm oil with hydrogenated coconut oil, cottonseed oil, hydrogenated cottonseed oil, and an American fat blend containing a mixture of butterfat, tallow, lard, shortening, salad oil, peanut oil and corn oil. At the end of a 14-month feeding period, coconut oil fed rabbits showed the most atherosclerotic lesions, while palm oil-fed rabbits showed less lesions compared to the coconut oil fed rabbits.

More than 70% of the vitamin A intake in Third World countries comes from fruits and vegetables in the form of carotenoids (Van Rooyen et al., 2008). In humans and animals, carotenoids, an important constituent of palm oil, play an important role in protection against photo-oxidative processes by acting as oxygen and peroxyl radical scavengers. Their synergistic action with other antioxidants makes them an even more potent compound.

It has been suggested that different individual compounds exhibiting a variety of anti-oxidant activities may provide additional protection against oxidative stress when ingested simultaneously (Esterbauer et al., 1991). A combination of lipophilic anti-oxidants present in red palm oil results in an inhibition of lipid peroxidation which is significantly greater than the sum of the individual effects (Zhang et al., 1995). This suggests that a cocktail of anti-oxidants may have a far more profound anti-oxidative effect due to the synergistic action of the individual compounds (Zhang et al., 1995).

The antioxidant properties of RPO has been attributed to the synergistic actions of carotenoids and vitamin E in the presence of lycopene in a natural food environment and this might provide the ultimate dietary supplement to fight disease associated with oxidative stress (Van Rooyen et al., 2008).

Conclusively, it could be said that oxidative stress plays a role in inflammatory and chronic diseases such as HIV/AIDS and TB and contribute significantly to depletion of immune factors, micronutrients and also promotes the progression of disease. Red palm oil could potentially retard the process due to its unique characteristics and also as it is known to be rich in several important antioxidants.

8. References

Allard JP, Aghdassi E, Chau J, Tam C, Koracs CM, Salit E & Walmsley SL (1998). Effect of vitamin E and C supplementation on oxidative stress and viral load in HIV-infected subjects. AIDS 12:1653-1659.

Barbosa KB, Bressan J, Zulet MA & Martínez-Hernández JA (2008). Influence of dietary intake on plasma biomarkers of oxidative stress in humans. Anales del Sistema Sanitario de Navarra 31(3): 259-80.

Bartlett JD (1998). Natural history and classification of HIV infection Chapter 2 In: Medical Management of Infection 1-15. Bartlett, JG (Ed.). Baltimore: Port City Press.

Berk, BC (2007). Novel approaches to treat oxidative stress and cardiovascular diseases. Trans Am Clin & Climatol Assoc 118: 209-14.

Bester DJ, Van Rooyen J, Du Toit EF & Esterhuyse AJ (2006). Red palm oil protects against the consequences of oxidative stress when supplemented with dislipidaemic diets. Med Tech SA 20: 3-8.

Bloom BR & Murray (1992). Tuberculosis: Commentary on a re-emergent killer. Sci 257: 1055-1064.

Bonanome A & Grundy SM (1988): Effect of dietary stearic acid and plasma cholesterol and lipoprotein levels. New Eng J Med 318:1124-1128.

Ceconi, C., Boraso, A., Cargnoni, A. & Ferrari, R. (2003). Oxidative stress in cardiovascular disease: myth or fact? Arch of Biochem & Biophys 420 (2): 217-21.

Champe PC, Harvey RA & Ferrier DR (2005). Biochemistry. RA Harvey & PC Champe (Eds) 3rd Ed Lippincott Williams & Wilkins, Ambler, PA, USA.

Choudhury N, Tan L & Truswell AS (1995): Comparison of palm olein and olive oil : Effects on plasma lipids and vitamin E in young adults. Am J Clin Nutr 61: 1043-1051

Cunningham-Rundles S (2001). Nutrition and the mucosal immune system. Current Opin Gastroenterol 17: 171-176.

Cunningham-Rundles S, McNeeley DF & Moon A (2005). Mechanisms of nutrient modulation of the immune response. J Allergy Clin Immunol 115 (6):1119-1128; quiz 1129.

Das UN, Podma M, Sogar PS, Ramesh G & Koratkar R (1990). Stimulation of free radical generation in human leucocytes by various agents including NF is a calmodulin – dependent process. Biochem Biophys Res Commun 67: 1030-6.

Di Massio P, Murphy M & Seis H (1991). Antioxidant defense systems: the role of carotenoids, tocopherols and thios. Am J Nutr 53: (suppl) 1945-2005.

Dye CS, Scheele P, Dulin V, Pathanira S and Raviglione MC.(1999). Global burden of tuberculosis, estimated incidence, prevalence and mortality. WHO. Global Surveillance and Monitoring Project, JAMA 282: 677-686.

Ebong PE, Owu DU & Isong EU.(1999). Influence of palm oil (Elaesis guineensis) on health. Plant Foods Hum Nutr 53:209-222.

Edem DO (2002). Palm oil: biochemical, physiological, nutritional, haematological and toxicological aspects of: a review. Plant Foods Hum Nutr 57: 319-41.

Engelbrecht AM, Esterhuyse AJ, du Toit EF. Lochner A & Van Rooyen J (2006). P38-MAPK and PKB/Akt, possible role players in red palm oil-induced protection of the isolated perfused rat heart? J Nutr Biochem 17: 265-271.

Enger CN, Graham J & Peng Y (1996). Survival from early, intermediate and late stages of HIV infection. J Am Med Assoc 275:1329-1334.

Esterbauer H, Schaur JS & Zollner H (1991). Chemistry and biochemistry of 4-hydroxynonenal, malondialdehyde and related aldehydes. Free Rad Biol Med 11:81-128.

Esterhuyse AJ, du Toit EF, Van Rooyen J (2005). Dietary red palm oil supplementation protects against the consequences of global ischemia in the isolated perfused rat heart. Asian Pac J Clin Nutr 14(4): 340-347 (a).

Farooq A; Amir ID & Syed AA (1996). Dietary fats and coronary risk factors modification-a human study. Proc PIPOC

Fortuño A, San José G, Moreno MU, Díez J & Zalba G (2005). Oxidative stress and vascular remodeling. Exp Physiol 90 (4): 457-66.

Fridovich I (1970). Quantitative aspects of the production of superoxide anion radical by milk xanthine oxidase. J Biol Chem 245:4053-4057.

Frieden TR, Fujiwara PI, Washko RM & Hamburg MA (1995). Tuberculosis in New York City-turning the tide. N Engl J Med 333:229-233.

Gebick JM & Bielski BHJ (1981). Comparison of the capacities of the dihydroxyl and superoxide radicals to initiate chain oxide of linoleic acid. J Am Chem Soc 103:7020-7022.

Gil L, Martinez G., Gonzalez I, Tarinas A, Alvarez A, Giuliani A, Molina R, Tapanes R, Perez J & Leon OS (2003). Contribution of characterization of oxidative stress in HIV/AIDS patients. Pharmacol Res 47 (3): 217 – 24.

Grossman Z & Heberman RB (1997). T-cell homeostasis in HIV infection is neither failing nor blind: Modified cell count reflects an adaptive response of the host. Nat Med 3: 486-490.

Haase AT (1999). Population biology of HIV-infection: viral load and CD4+T-cell demographics and dynamics in lymphatic tissues. Ann Rev Immunol 17: 625-656.

Haber F & Weiss J (1994). The catalyzed decomposition of hydrogen peroxide by iron salts. Proc Royal Soc A 147-332.

Hamilton CA, Miller WH, Al-Benna S, Brosnan MJ, Drummond RD, McBride MW & Dominiczak AF (2004). Strategies to reduce oxidative stress in cardiovascular disease. Clin Sci 106 (3): 219-34.

Hayes KC, Pronczuk A & Khosla P (1995). A rationale for plasma cholesterol modulation by dietary fatty acids: Modelling the human response in animals. J Nutr Biochem 6:188-194.

Hayes KC (1993). Specific dietary fat without reducing saturated fatty acids does not significantly lower plasma cholesterol concentrations in normal males. Am J Clin Nutr 55:675-681.

Hayes, KC, Pronczuk A, Lindsey S & Diersen-Scchade D (1991). Dietary saturated fatty acids (12:0,14:0,16:0) differ in their impact on plasma cholesterol and lipoproteins in human primates. Am J Clin Nutr 53:491-498.

Heistad DD, Wakisaka Y, Miller J, Chu Y & Pena-Silva R (2009). Novel aspects of oxidative stress in cardiovascular diseases. Cir J 73 (2): 201-07.

Higashi Y, Jitsuiki D, Chayama K & Yoshizumi M (2006). Edaravone (3-methyl-1-phenyl-2-pyrazolin-5-one), a novel free radical scavenger, for treatment of cardiovascular diseases. Recent Pat Cardio Drug Discov 1 (1):85-93.

Holub BJ, Silicilia F & Mahadevappa VG (1989). Effect of tocotrienol derivatives on collagen and ADP-induced human platelet aggregation. Presented at PORIM. Int Palm Oil Dev Conf, Sept 5-6, Kuala Lumpur.

Hornstra G (1988). Dietary lipids and cardiovascular disease: Effects of palm oil. Oleagineux 43:75-8.1

Jagirdar J & Zagzag D (1996). Pathology and insights into pathogenesis of TB. pp. 467-482 in Tuberculosis 1st Ed. Little Brown & Company. New York N. Y.

Januga P, Jaruga B, Gackowski D, Olczak A, Halota W, Pawlowska G & Olinski R (2002). Supplementation with antioxidant vitamins prevent oxidative modification of DNA in lymphocytes of HIV infected patients. Free Rad Biol Med 32: 414-420.

Khosla P & Hayes KC (1992). Comparison between effects of dietary saturated (16:0), monounsaturated (18:1) and polyunsaturated (18:2) fatty acids on plasma lipoprotein metabolism in Cebus and Rhesus monkeys fed cholesterol-free diets. Am J Clin Nutr 55:51-62.

Khosla P & Hayes KC (1994). Cholesterolaemic effects of the saturated fatty acids of palm oil. Food Nutr Bull 15:119-125.

Kiedzierska K & Crowe SM (2001). Cytokines and HIV: Interaction and clinical implications. Antivir Chem Chemother 12: 133 – 150.

Kiremidjian-Schumacher L, Roy M, Wishe HI, Cohen MW & Stotzky G (1994). Supplementation with selenium and human immune cell functions: Effect on cytotoxic lymphocytes and natural killer cells. Biol Trace Elem Res. 41 (1-2):115-127.

Koch R (1882). Die aetiology der Tuberculese. Klinische Wochenschr. 19:221-230.

Koot M, Van't B, Wout A & Koot NA (1996). Relation between changes in cellular load evaluation of viral phenotype and the clonal composition of virus type infection. J Infect Dis 173:349-354.

Kooyenga DK, Gerler M, Watkins TR, Gapor A, Diakoumakis E & Bierenbaum ML (1997). Palm oil antioxidant effects in patients with hyperlipidaemia and carotid stenosis-2 year experience. Asia Pac J Clin Nutr.6:1:72-75.

Krichevsky D (1988). Effects of triglyceride structure and lipid metabolism. J Nutr Rev 46:177.

Kritchevsky D (2000). Impact of red palm oil on human nutrition and health. Food & Nutr Bull, 21: 182-88.

Kritchevsky D (1996)). Influence of triglyceride structure on experimental atherosclerosis in rabbits. FASEB J 10:A187.

Kurfeld D, Davidson LM & Lopex-Guisa JM (1990). Palm and other edible oils: atherosclerosis study in rabbits. FASEB 4: 23-26.

Lindquist Y, Branden CL, Mathews FS & Lederer F (1991). Spinach glycolate oxidase and yeast flagrant cytochrome are structurally homologous and evolutionary related enzymes with distinctly different function and flavin mononucleotive binding. J Biol Chem 266:3198-3207.

Loschen G, Azzi S, Richter C & Flohebp L (1974). Superoxide radicals as precursors of mitochondrial hydrogen peroxide. FASEB 42: 68-72.

Madebo T, Lindtjϕrn B, Aukrust P & Berge R (2003). Circulating anti-oxidants and lipid peroxidation products in untreated tuberculosis patients in Ethiopia. Am J Clin Nutr 78: 117 – 22.

Marzuki A, Arshad F, Tariq AR & Kamsiah J (1991). Influence of dietary fat on plasma lipid profiles of Malaysian adolescents. Am J Clin Nutr 53:1010S-1014S.

Mathema B, Kurepina NE, Bifani PJ & Kreiswirth BN (2006). Molecular epidemiology of tuberculosis: Current insight. Clin Microbiol Rev 19: 658-685.

Mattson FH & Grundy SM (1985). Comparison of effects of dietary saturated, monounsaturated and polyunsaturated fatty acids on plasma lipids and lipoproteins in man. J Lipid Res 26:194-202.

Maxwell SR & Lip GY (1997). Free radicals and antioxidants in cardiovascular disease. Brit J Clin Pharmacol 44 (4): 307-17.

Mclntosch GH, Whyte J. McAthur R & Nestle PJ (1991). Barley and wheat foods; Influence on plasma cholesterol concentrations in hypercholesterolaemic men. Am J Clin Nutr 1205-1209.

Medlar EM (1955) Necropsy studies of human pulmonary TB. Ann Rev Tuberc 71 (2): 29 – 55.

Murray RK (2000). Muscle and the cytoskeleton. In Harper's Biochemistry (Murray, R. K., Granner, D. K.,Mayes, P. A., and Rodwell, V. W. Eds., 25th ed.), pp. 715-736. New York, McGraw-Hill Press.

Ng TKW, Hayes KC, de Witt GF, Jegathesan M, Satgunasingham N, Ong, ASH & Tan DTS (1992). Palmitic and oleic acids exert similar effects on serum lipid profile in the normo-cholesterolaemic humans. J Am Coll Nutr 11:383-390.

Ng TKW, Hassan K, Lim JB, Lye, MS & Ishak R (1991). Non-hypercholesterolaemic effects of a palm oil diet in Malaysian volunteers. Am J Clin Nutr 53:1015S-1020S.

O'Brien RJ (2001). Tuberculosis: scientific blueprint for TB drug development.Global Alliance for TB. Drug Dev, New York, N.Y.

Oguntibeju OO, van den Heever WMJ & Van Schalkwyk FE (2008). Potential effects of nutrient supplement on the anthropometric profiles of HIV-positive patients: complementary medicine could have a role in the management of HIV/AIDS. Afr J Biomed Res 11: 13-22.

Oguntibeju OO, van der Heever WMJ & Van Schalkwyk FE (2007a). Interplay between socio-demographic variables, nutritional and immune status of HIV-positive/AIDS patients. Pak J Biol Sci 10 (20): 3592 – 98.

Oguntibeju OO, van der Heever WMJ & Van Schalkwyk FE (2007b). A review of the epidemiology, biology and pathogenesis of HIV. J Biol Sci 7(8):1296-1304.

Ong ASG & Goh SH (2002). A healthful and cost-effective dietary component. United Nat Food & Nutr Bull 23 (1): 11-19.

Paxon WA, Martin SR & Tse D (1996). Relative resistance to HIV-1 infection of CD4 lymphocytes from persons who remain uninfected despite multiple sexual exposures. Nat Med 2: 412-417.

Qureshi A, Qureshi N & Hasler-Rapaczl (1991a). Dietary tocotrienol reduces concentrations of plasma cholesterol, apoprotein B, thromboxane B2 and platelet factor 4 in pigs with inherited hyper-lipidemias. Am J Clin Nutr 53:1042S-1046S.

Qureshi A, Burger WC, Prentice N, Bird HR & Sunde ML (1980). Regulation of Lipid metabolism in chicken liver by dietary cereals. J Nutr 10:388-393.

Qureshi A, Qureshi N & Wright JJK (1991b): Lowering of serum cholesterol in hypercholesterolemic humans by tocotrienols (palmvitee). Am J Clin Nutr 53:1021S-1026S.

Reddy YN, Murthy SV, Krishna DR & Prabhakar S (2004). Role of radicals and antioxidants in tuberculosis patients. Ind J Tuberc 151: 213-18.

Roach DR., Bean AG, Demangel C, Frances MP, Briscoe H & Brilton WJ (2002). TNF regulates chemokine induction essential for cell recruitment, granuloma formation and clearance of mycobacterium infection. J Immunol 168: 4620 – 27.

Sebinova E, Kagan V, Han D & Parker L (1991). Free radical recycling and inter-membrance mobility in the antioxidant properties of alpha– tocopherol and beta– tocotrienol. Free Rad Biol Med 10: 263 – 275.

Semba RD & Tang L (1999).   Micronutrients and the pathogenesis of human immunodeficiency virus infection. Br J Nutr 81: 181-189.

Stephen MS (2006). The pathogenesis of HIV infection: the stupid may not be so dumb afterall. Retrov 3: 60.

Sundram K, Ismail A, Hayes KC, Jeyamalar R & Pathmanathan R (1997). Trans-celeidial fatty acids adversely affect the lipoprotein profile relative to specific saturated fatty acids in humans. J Nutr 127:3:5143-5203.

Sundram K, Sambanthamurthi R, Tan YA (2003). Palm Fruit Chemistry. Asia Pac J Clin Nutr 12 (3):` 355 – 362.

Sundram K, Hornstra G & Houwelingen AC (1992). Replacement of dietry fat with palm oil: Effect on human serum lipid, lipropotein and apoprotein. Brit J Nutr 68:677-692.

Sundram K, Khor HT & Ong ASH (1990). Effect of dietary palm oil and its fractions on rat plasma LDL and high density lipoprotein. Lipids 25(4):187-193.

Truswell AS, Choudhury N & Roberts DCK (1992). Double-blind comparison of plasma lipids   in healthy subjects eating potato crisps fried in palm olein or canola oil. Nutr Res 12: S34-S52.

Valko M, Leibfritz D, Moncol J, Cronin MT, Mazur M. & Telser J (2007). Free radicals and antioxidants in normal physiological functions and human disease. Int J Biochem Cell Biol 39 (1): 44-84.

Van Rooyen J, Esterhuyse AJ, Engelbrecht A & du Toit EF (2008). Health benefits of natural carotenoid rich oil: a proposed mechanism of protection against ischaemia reperfusion injury. Asia Pac J Clin Nutr 17: 316 – 19.

Waterfall AH, Singh G, Fry JR & Marsden CA (1997). The measurement of lipid peroxidation *in vivo*. Brain Res Protocols 2: 217–22.

Weiss R (1996). HIV receptors and the pathogenesis of AIDS. Sci., 272: 1885-1886.

Zalba G, Fortuño A & Díez J (2006). Oxidative stress and atherosclerosis in early chronic kidney disease. Nephrol Dialy Transplant 21 (10): 2686-90.

Zhang J, Wang CR, Dai J, Chen XS & Ge KY (1997a): Palm oil diet may benefit mildly hyper-cholesterolaemic Chinese adults. Asia Pac J Clin Nutr 6:1:22-25.

Zhang J, Wang P, Wang CR Chen XS & Ge KY (1997b): Non-hypercholesterolaemic effects of a palm oil diet in Chinese adults. J Nutr 127(3): 5095-5135.

Zhang J, Wang P, Wang P, Wang CR, Chen XS & Ge KY (1995). The effects of palm oil on serum lipids and thrombosis in Chinese adults. Paper presented at the 7th Asian Congr Nutr, / Beijing, China Oct. 7-11.

8

In Vitro Leukocyte Adhesion in Endothelial Tissue Culture Models Under Flow

Scott Cooper, Melissa Dick, Alexander Emmott,
Paul Jonak, Léonie Rouleau and Richard L. Leask
Department of Chemical Engineering, McGill University,
Canada

1. Introduction

Atherosclerosis, an inflammatory disease which causes thickening and stiffening of arteries, is a major cause of death in the United States (Lloyd-Jones et al., 2010). These deaths occur because of vessel occlusion created by atherosclerotic plaques and thrombus shedding, leading to heart attack, ischemia, or stroke. Atherosclerosis is expected to be the leading cause of death worldwide within 10 years (Lloyd-Jones et al., 2010).

Inflammation plays a significant role in the initiation and progression of atherosclerosis. The cells that line the arteries, endothelial cells (ECs), mediate the inflammatory process. The forces created by blood flow affect the inflammatory response of ECs and the interaction with blood components. This chapter summarizes the background in our recent studies on the response of ECs to blood forces and the interaction of inflammatory cells.

2. Background

2.1 Pathogenesis of atherosclerosis

Vascular anatomy

Arteries have three tissue layers: the intima, media, and adventitia. The intima is lined with a monolayer of ECs in direct contact with blood. The ECs act as a protective membrane, allowing diffusion from the blood stream into the artery. ECs are capable of expressing specific genes in response to physical stresses which cause the vessel to remodel leading to the development of atherosclerosis. In addition, the intima can contain other cells (smooth muscle cells, fibroblasts and inflammatory cells), an extracellular matrix (ECM), and is only a few cell layers thick in healthy tissue. The internal elastic membrane, consisting of a layer of elastic connective tissues, separates the intima and media. The media layer is mainly comprised of smooth muscle cells (SMCs) and ECM. Although the media is involved in atherosclerosis, remodelling is less evident. The adventitia is relatively unaffected by atherosclerosis. It is separated from the media by the external elastic membrane comprised mainly of collagen, providing structural support yet allowing for artery expansion when required (Waller et al., 1992).

Over time, an atherosclerotic plaque grows by the accumulation of lipids, inflammatory cells, vascular cells and matrix material in the intima. It often produces a fibrous cap, over a necrotic lipid core, which can weather and rupture over time. The artery is able to compensate for some intimal thickening by expanding outwards, instead of allowing for the plaque to impede blood flow. Eventually the vessel can no longer expand outwards, and negative remodelling can occur. Blood flow is therefore disturbed through the formation of a stenosis (Shah, 2006). This may lead to ischemia and angina pectoris (Libby, 2002).

Atherosclerosis can occur in any size of artery. However, clinical manifestations frequently occur in medium and large arteries when the EC layer is breached by erosion or disruption of the fibrous cap. Disruption may occur from a thinning of the fibrous cap as there is increased lipid accumulation, inflammatory cell recruitment and matrix metalloproteinase (MMP) expression, as well as the expression of cytokines inhibiting collagen synthesis. When the plaque is opened up to the blood stream, platelets cause blood coagulation and thrombus formation. There are two possible outcomes after thrombus formation. First, the thrombus may be broken down and reabsorbed. A second outcome is that the thrombus is disrupted by the blood flow and detached from the site of injury. This embolism may then travel through the vasculature to small arteries, where it causes ischemia and potential heart attack or stroke (Libby, 2002).

The role of inflammation

Over the past two decades, it has been recognized that inflammation plays a critical role in the development and progression of atherosclerosis (Libby, 2002). Indeed, the localization of plaques to regions of disturbed blood flow (curvature, bifurcations, and branches) has been linked to an inflammatory response of ECs due to hemodynamic forces (Libby, 2002; Shah, 2006). In these areas, ECs become inflamed, causing an influx of leukocytes (Shah, 2006). It has been found that nuclear factor κB (NF-κB), a transcription factor responsible for expressing genes involved in the inflammatory cascade, is activated at sites of disturbed blood flow (Van der Heiden et al., 2010).

Additionally, monocytes, part of the family of leukocytes, are attracted into the intima through the EC layer due to the existence of a chemical gradient. During inflammation, a chemokine called monocyte chemoattractant protein-1 (MCP-1) is expressed within the intima layer (Libby, 2002). MCP-1 is expressed constitutively, by both the EC layer, and the SMCs within the intima (Schwartz et al., 1991). The receptor for MCP-1 on the monocyte (the CCR2 receptor) is attracted to the MCP-1 within the intima, and monocytes migrate into the intima through diapedesis (Libby, 2002).

Also flowing in the blood stream are low-density lipoproteins (LDL), including cholesterol. LDLs are brought across the EC layer and into the intima. Reactive oxygen species within the intima, including OH and O_2, oxidize the LDLs, turning them into oxidized low-density lipoproteins (Ox-LDL). These Ox-LDL molecules are also responsible for stimulating ECs and SMCs to secrete additional MCP-1 (Schwartz et al., 1991).

Once inside the intima, monocytes begin to express characteristics of macrophages, activated by the presence of macrophage colony stimulating factor (M-CSF) (Libby, 2002). M-CSF also activates Ox-LDL receptors on the macrophages, turning them into scavengers for Ox-LDL (Libby et al., 2002). Macrophages begin to take up Ox-LDL, filling themselves with lipids and transforming into foam cells (Ross, 1993; Shah, 2006). The accumulation of macrophage foam cells, as well as collagen, elastin, and proteoglycans, within the intimal layer is known as the fatty streak, and is an early indication of a complex atherosclerotic lesion (Ross, 1993; Schwartz et al., 1991). It has been shown that the progression of a foam cell to a more advanced lesion may be halted or reversed, possibly through a decrease in blood LDL levels (Schwartz et al., 1991).

In advanced lesions, macrophages are unable to take up any additional Ox-LDL, and these lipid molecules begin to accumulate within the intima instead (Schwartz et al., 1991). Ox-LDLs are toxic, and begin to injure and kill ECs, SMCs, and macrophages. When the macrophage is injured, its lipid contents are released into the intimal layer, forming a lipid core, also comprised of enzymes, cytokines, and growth factors that have accumulated within the intima (Schwartz et al., 1991; Shah, 2006).

In order to protect the body from the growing necrotic lipid core within the intima, a fibrous plaque is formed over the lesion. This prevents direct contact between the accumulation of cells within the intima and blood flow. The fibrous plaque is composed primarily of collagen, elastin, and proteoglycans that were found in the original fatty streak (Ross, 1999). T lymphocytes and macrophages release MMPs, which break down the extracellular matrix within the intima, allowing for these components to be used within the fibrous plaque (Libby, 2002).

2.2 Mechanisms in leukocyte-endothelium adhesion
2.2.1 Intercellular adhesion at the arterial surface
The initiation of atherosclerosis is hypothesized to start with endothelial injury, which triggers inflammatory pathways integral to the progression of the disease (Ross et al., 1977). Leukocytes, including neutrophils and monocytes, preferentially adhere to sites of inflammation. The events that take place during leukocyte recruitment are shown in the representative drawing, Figure 1.

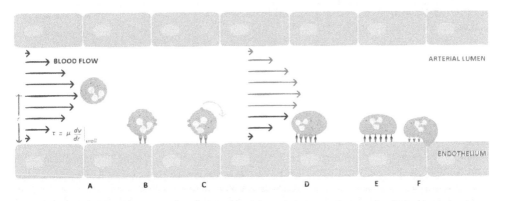

Fig. 1. Leukocyte recruitment to a site of endothelial injury. When injured, the endothelium expresses an increase in cell adhesion molecules (CAMs). (A) Leukocytes circulate within the blood stream. (B) Ligands on the leukocytes attach to selectins on the endothelium, effectively tethering the leukocyte. (C) The leukocyte begins to roll across the endothelium, reducing its velocity by forming and breaking selectin-ligand bonds. (D) Leukocytes begin to make bonds between CAMs and integrins. This firmly attaches the leukocyte to the endothelium. (E) Leukocytes migrate towards a cellular junction through CAM-integrin interactions. (F) CAM-CAM interactions allow the leukocyte to migrate through the endothelium by diapedesis.

Leukocyte tethering and rolling

Leukocytes circulating within the bloodstream must make their way to the site of injury, located on the endothelium. Although leukocytes flow in close contact with the endothelial layer, they do not stick until the inflamed endothelium starts to express adhesive molecules (Kelly et al., 2007). Tethering and rolling of leukocytes along the endothelial wall is due to a class of cell adhesion molecules (CAMs) known as selectins. There are three selectins involved: E-selectin, P-selectin, and L-selectin. Both E- and P-selectins are expressed on the endothelium; E-selectin is synthesized and expressed after endothelial stimulation, whereas P-selectin is expressed constitutively and stored, then quickly released upon stimulation (Kelly et al., 2007). L-selectin differs in that it is not expressed on the endothelium, but instead is constitutively expressed on the leukocyte surface. Both E- and P-selectin recognize carbohydrate ligands on the surface of leukocytes, while L-selectin recognizes a series of ligands expressed on the endothelium. When the selectins come in contact with their ligands they will bind, tethering the leukocyte to the endothelium (Miyasaka et al., 1997). Tethering facilitates leukocyte rolling along the EC surface. The velocity of the travelling leukocyte will be reduced as more selectin-ligand bonds form, allowing leukocyte adhesion (Kelly et al., 2007; Kubes & Kerfoot, 2001).

Leukocyte adhesion

Firm adhesion begins when integrins, another class of CAMs, are activated on the surface of leukocytes by chemoattractant cytokines, termed chemokines. Chemokines are secreted by circulating leukocytes and ECs (Kelly et al., 2007). One class of integrin responsible for neutrophil adhesion is the β_2-integrin. When activated by cytokines, certain β_2-integrins bind to intracellular adhesion molecules 1 and 2 (ICAM-1 and ICAM-2, respectively) on the EC surface, effectively adhering the leukocyte to the EC. Also activated by cytokines, $\alpha_4\beta_1$-integrin will bind with vascular cell adhesion molecule 1 (VCAM-1) (Kelly et al., 2007; Rao et al., 2007). The bound integrin-CAM complex results in firm adhesion of the leukocyte to the endothelial surface (Miyasaka et al., 1997). Although under transcriptional regulation, both adhesion molecules ICAM-1 and VCAM-1 have been shown to be upregulated at sites in the vasculature prone to developing atherosclerotic lesions (Iiyama et al., 1999).

Leukocyte migration

Once the leukocytes have firmly adhered to the endothelium, they may migrate from the lumen of the blood vessel into the subintimal space. Initially, leukocytes must make their way to the closest EC junction through a process termed locomotion. The movement of the leukocyte is made possible through interactions with leukocyte integrins and both ICAM-1 and -2 located on the endothelial surface (Schenkel et al., 2004). At the junction, the leukocytes will encounter another cell adhesion molecule, called platelet endothelial CAM 1 (PECAM-1). PECAM-1 is expressed both on the leukocyte and the endothelium. An interaction between the complementary PECAM-1 molecules allows leukocytes to migrate through the gap junction by diapedesis (Rao et al., 2007; Schenkel et al., 2004), a process also called transmigration.

2.2.2 Idealized arterial hemodynamics

Hemodynamics, the mechanics of blood flow, influence many of the physiological processes of the vascular system (Glagov et al., 1988). From an engineering perspective, blood flow through medium and small arteries (such as the right and left coronary arteries) is often simplified by assuming steady laminar flow in a straight, rigid vessel (Ku, 1997; Nichols & O'Rourke, 1990). Additionally, blood is assumed to be a Newtonian fluid to simplify the flow dynamics to Hagen-Poiseuille flow (Nichols & O'Rourke, 1990). Such assumptions allow us to reduce the governing equations describing pressure-driven flow into a one-dimensional velocity profile in the form of Hagen-Poiseuille flow (Nichols & O'Rourke, 1990). For a cylindrical vessel model of arterial perfusion, the velocity profile as a function of a radial dimension is described by:

$$v(r) = 2 \left(\frac{Q}{\pi R^2} \right) \left[1 - \left(\frac{r}{R} \right)^2 \right] \tag{1}$$

where Q is the volumetric flow rate of blood and R is the hydraulic radius of the vessel.
Wall shear stress (WSS) is a tangential force per unit area of a fluid-wall interface that results from flow parallel to the vessel wall. For fluids with constant dynamic viscosity μ (Newtonian fluid), the WSS is the product of the viscosity and the shear rate γ, evaluated at the vessel wall:

$$\tau_w = \mu \gamma |_{r=R} \tag{2}$$

The wall shear rate of a fluid is the velocity gradient evaluated at the fluid-wall interface. For Hagen-Poiseuille flow through a cylindrical vessel (Equation 1), the wall shear rate is expressed as:

$$\gamma |_{r=R} = \left. \frac{dv(r)}{dr} \right|_{r=R} = \frac{4Q}{\pi R^3} \tag{3}$$

Combining Equations (2) and (3) produces an expression for WSS that is dependent on both the vessel geometry and the volumetric flow rate:

$$\tau_w = \mu \frac{4Q}{\pi R^3} \tag{4}$$

This equation is accepted as a reasonable model of the average WSS for arteries that are absent from serious geometric disturbances (Ku, 1997). Arterial WSS values range from 5 to 70 dyne/cm² with average WSS values of approximately 15 dyne/cm² being observed in coronary arteries (Glagov et al., 1988; Malek et al., 1999). Moderate levels of steady, laminar shear stress (> 10-15 dyne/cm²) are believed to induce an atheroprotective EC phenotype while low shear stresses (< 4 dyne/cm²) are believed to induce an atheroprone EC phenotype (Malek et al., 1999). An atheroprone EC phenotype describes one which facilitates the disease pathway marked by an increase in adhesion molecules and a decrease in vasodilators (as described in Sections 2.1 & 2.2) (Libby, 2002; Malek et al., 1999).
The fluid flow regime is determined by the dimensionless Reynolds number (Re), which represents the ratio of inertial to viscous forces. For flow in a cylindrical channel, the Reynolds Number is described by:

$$Re = \frac{\rho D U}{\mu} \tag{5}$$

where ρ is the fluid density, D is the hydraulic diameter, and U is the average fluid velocity. A Reynolds Number below 2300 indicates laminar flow that will behave predictably while a value above this threshold suggests the presence of flow disturbances. The average arterial conditions are within a laminar flow regime (Nichols & O'Rourke, 1990) and vary depending on the artery and metabolic demand (Myers et al., 2001; Nichols & O'Rourke, 1990).

Localized hemodynamics of leukocyte adhesion

The progression of leukocyte adhesion is strongly influenced by local hemodynamic forces. As the cell is passing along the wall, the torque imparted on the cell by the blood stream causes the cell to spin. As a result, the state of loose attachment with selectin-ligand bonds constantly forming and breaking has become known as cell rolling. The blood stream imposes not only torque but also shear stress on the slow moving cell. In turn, the membrane of the cell will try to distribute this stress by elongating in the direction of flow, allowing for increased binding with the vessel wall. Firrell and Lipowsky found that leukocytes rolling along rat arteriolar walls would elongate by around 140%, allowing their contact area to jump from approximately 14 μm² to 50 μm² (Firrell & Lipowsky, 1989).

Modelling of bond forces and leukocyte attachment is well documented in the literature (Cozens-Roberts et al., 1990; Evans et al., 2004; Lawrence et al., 1997; Tees & Goetz, 2003). For successful adhesion, a fine balance between the adhesive force and the hemodynamic force must be met. This adhesive force is dependent on several factors, including: the receptor density of both cells, the rate of reaction with respect to both bond formation and dissociation and the strength of the bonds and their response to strain. For instance, the bonds between E-, P-, and L-selectins and their respective ligands behave as catch-slip bonds (Lawrence et al., 1997; Marshall et al., 2003; Sarangapani et al., 2004). Though receptor-ligand bonds spontaneously dissociate (Tees & Goetz, 2003), slip bond behaviour describes an increasing probability of dissociation with increasing tensile force. Catch-slip bond behaviour, on the other hand, strengthens with increasing tensile force until some optimal force has been met. Furthermore, the rate at which this force is increased, known as the force gradient, also affects the strength of the selectin-ligand bonds (Evans et al., 2004). This is of particular interest to the study of stenotic arteries as plaque formation leads to distinct regions of varying force gradients.

2.3 Leukocyte adhesion in Parallel-Plate Flow Chambers (PPFCs)

The flow between two parallel plates has often been used to investigate the effects of blood flow on ECs and their interactions with blood components. Traditionally, parallel-plate flow chambers (PPFCs) have been used to provide an environment suited for tissue and suspension culture experiments under laminar flow. In classical PPFCs, fluid is driven through a channel formed by two narrowly separated plates in parallel. ECs are cultured on the bottom surface of the upper plate (often, a glass coverslip) while suspension cultures of leukocytes (neutrophils or monocytes) are prepared in the perfusion medium and their movement visualized within the chamber (Lawrence et al., 1987). This allows for regional or complete surface quantification of cells that are either adherent or, if observed in real-time, leukocytes that are undergoing rolling adhesion. A schematic of a PPFC is presented in Figure 2.

In a well defined PPFC, the WSS can be accurately characterised for steady, laminar flow of a Newtonian fluid as a function of a constant measurable volumetric flow rate Q:

$$\tau_w = \frac{6\mu Q}{wh^2} \qquad (6)$$

where w is the width of the plate perpendicular to flow and h is the height of the interstitial gap between the two plates. For a constant volumetric flow rate, the velocity profile is parabolic and the WSS is uniform across the upper and lower plates save for the boundaries defined by the gasket (where the flow field approaches zero) and in the region of developing flow at the inlet of the chamber (Bacabac et al., 2005; Lawrence et al., 1987). In practice, PPFCs are designed with a large w/h ratio allowing most of the flow field to be homogenous over the surface of the cells (Bacabac et al., 2005).

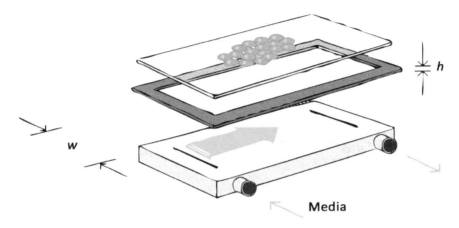

Fig. 2. Schematic of a Parallel-Plate Flow Chamber.

Studies of leukocyte adhesion using PPFCs have become the benchmark for revealing the role of shear in the adhesion pathway. Early results revealed a discreet shear dependence on both non-specific (Forrester & Lackie, 1984) and adhesion molecule-mediated adhesion (Alon et al., 1995; Finger et al., 1996; Lawrence et al., 1997). Further studies highlighted the role of endothelial dysfunction and inflammation as a precursor to adhesion when conditioned with flow (Alcaide et al., 2009; Sheikh et al., 2003; Sheikh et al., 2005). These findings complement the paradigm of leukocyte adhesion at sites of vascular inflammation; however, they do not address focal adhesion in non-uniform shear fields. This was considered by performing flow experiments using a step disturbance across the plate of the flow chamber (Burns & DePaola, 2005; Chen et al., 2006). The flow fields created by the step introduce regions of flow reversal and spatial WSS gradients to represent physiological hemodynamics (Burns & DePaola, 2005; Chen et al., 2006). Despite disturbed flow, leukocyte adhesion is increased in areas of high WSS gradients, with the highest incidence in re-attachment zones (Chen et al., 2006).

3. Asymmetric stenosis tissue culture model

3.1 Experimental methods
3.1.1 Asymmetric stenosis model design
Parallel-plate flow chambers are not ubiquitous when characterizing the role of WSS in endothelial dysfunction and leukocyte adhesion. Cone-plate viscometers (Shankaran & Neelamegham, 2001), animal models (Walpola et al., 1993, 1995) and three-dimensional (3D) tissue culture models (Hinds et al., 2001) have an increasing presence in the field.

A 3D model of an idealized coronary artery with an eccentric stenosis has been developed by our research group to reveal the effect of spatial WSS gradients on both endothelial inflammation and leukocyte adhesion (Rouleau et al., 2010a, 2010b). The eccentric stenosis geometry with a 50% occlusion (i.e. 50% area reduction, orthogonal to flow) has been chosen to represent a clinically relevant atherosclerotic lesion (Brunette et al., 2008; Wexler et al., 1996). The model measures 10 cm in length with an internal diameter of 3.175 mm, Figure 3. Sylgard™ 184 silicone elastomer is cast and cured in PVC moulds to create semi-compliant structures that maintain their geometric integrity through the stages of sterilization and cell culture preparation.

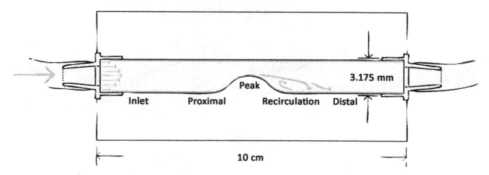

Fig. 3. Three-dimensional asymmetric stenosis model schematic with regional classifications defined using computational fluid dynamics and photochromic molecular flow visualization (Section 3.1.3).

3.1.2 Perfusion design
The perfusion flow loop consists of a media reservoir with tubing, flow dampeners, and an 8-roller peristaltic pump head with a programmable drive to produce steady or pulsatile laminar flow at the entrance of the models, Figure 4. ECs are cultured in the internal lumen of the model until they form a continuous, confluent monolayer. Inlet WSS values of 4.5, 9 and 18 dynes/cm² were chosen to represent moderate physiological shear in coronary arteries whereas inlet WSS values of 1.25 and 6.25 dynes/cm² were chosen to represent moderate to high shear for *in vitro* neutrophil adhesion. The WSS field was experimentally and numerically determined (Rouleau et al., 2010b).

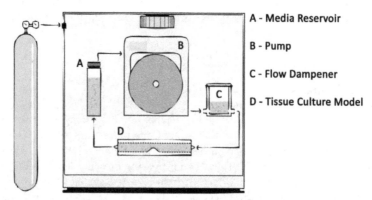

Fig. 4. Schematic of a steady-flow perfusion experiment with an asymmetric stenosis model.

3.1.3 Computational fluid dynamics and photochromic molecular flow visualization

Computational fluid dynamics (CFD) is a theoretical branch of research that relies on the power of modern computers to estimate fluid behaviour. The popularity of CFD lies in its ability to simulate physical experiments, thereby providing direction for further work or allowing for quick testing of key variables. With respect to cardiovascular flow studies, CFD is a numerical solution to a continuum of the Navier-Stokes (NS) equation. Although CFD is a powerful tool, it is only as accurate as the input data (e.g. geometry and mechanical properties). Defining the geometry and mechanics of healthy and diseased tissue is a constant endeavour in biomedical engineering (Choudhury et al., 2009; Tremblay et al., 2010). If the proper information is available, then CFD simulations are feasible, however experimental validation is still crucial. CFD has been performed for our stenosis model, yielding flow profiles at 6.25 dynes/cm^2 and 1.25 dynes/cm^2, respectively (Figures 5 & 6).

Fig. 5. Velocity profile, with an appreciable recirculation zone, in the asymmetric tissue culture model at 6.25 dynes/cm^2.

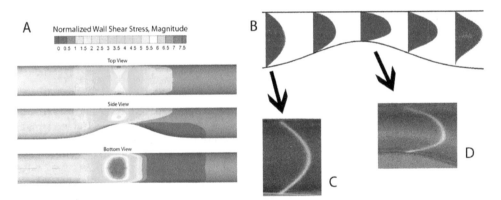

Fig. 6. Flow analysis of the asymmetric tissue culture model. (A) CFD normalized WSS contour plot. (B) Velocity profile at 1.25 dynes/cm^2. (C,D) PMFV velocity profiles at the inlet and peak, respectively.

We have used the photochromic molecular flow visualization (PMFV) technique to validate our CFD flow in the stenosis model (Ethier C.R. et al., 2000; Mahinpey et al., 2004). Using the flow profile at a given position, we are able to estimate the shear stress acting on the wall of a channel. Photochromic species reversibly change conformation when excited by a light source, such as a laser, resulting in an observable colour change. A PMFV setup includes a solution of photochromic dye, a laser, and a high-resolution camera. In practice, the photochromic dye solution is pumped through a micro-channel, the laser is triggered and the resulting pulse passes through the solution orthogonal to flow, activating any dye it contacts (Couch et al.,

1996; Park et al., 1999). At this moment, a narrow column of visible dye will appear within the solution. The solution is in motion, however, so a fraction of a second later, the excited dye will have displaced with flow. This displacement is recorded by a camera, providing a snapshot of the flow profile and subsequently, the shear stress on the opposing walls. Photochromic visualization results are presented in Figure 6 (c) & (d) to validate our CFD simulation.

3.1.4 Cellular analysis

EC morphology is often a good predictor of EC phenotype. Healthy ECs elongate in the direction of flow, whereas dysfunctional ECs may become randomly oriented and cobblestone in appearance (Dartsch & Betz, 1989). The shape index (SI) is a metric of EC morphology defined as (Nerem et al., 1981):

$$SI = \frac{4\pi \cdot Area}{Perimeter^2} \tag{7}$$

Generally, elongated cells have a lower SI than rounded cells. In concert with the SI, the angle of orientation evaluates the proportion of EC elongation relative to the direction of flow. ECs in regions of observed elongation will be narrowly distributed near 0° (i.e. in the axis of flow) while regions that appear cobblestoned will have a much wider distribution.

Protein and mRNA regulation of inflammatory markers and transcription factors defines the endothelial phenotype and relates to leukocyte adhesion and atherogenesis. Gene regulation is quantified for large cultures using Q-PCR, while the resultant protein expression is observed using Western Blotting. Regional inflammation around the stenosis can be observed using immunostaining and confocal microscopy for adhesion molecules and the translocation of inflammatory transcription factors to EC nuclei. Our analysis includes, but is not limited to, specific inflammatory markers and adhesion molecules, including: ICAM-1, VCAM-1, E-selectin and NF-κB.

3.2 Results and discussion
3.2.1 Endothelial cell morphology

The stenosis model was first used to investigate the morphological effects of shear gradients caused by the stenosis. A morphological response is one of the last measurable changes which occurs in the cascade of events following introduction of flow. As ECs experience a steady WSS, they tend to become elongated and aligned in the direction of flow, representing a healthy endothelium. When exposed to low shear magnitude or WSS spatial gradients, ECs tend to take on a more cobblestone and random morphology which is indicative of an unhealthy endothelium (Helmlinger et al., 1991; Levesque et al., 1986; Levesque & Nerem, 1985; Nerem et al., 1981; Nerem, 1993).

Perfusion experiments to evaluate EC response were run at wall shear stress values of 4.5, 9 and 18 dyne/cm² which corresponded to Reynold's numbers of 50, 100 and 200, respectively. It was found that at all times and inlet WSS values the shape indices in the inlet and outlet of the stenosis model were statistically similar to that of a straight model. These results demonstrate that these values can be a good reference point for morphological changes in the regions surrounding the stenosis. Furthermore, the longer the perfusion time, the more elongation was observed in the direction of flow.

Effect of wall shear stress gradients on endothelial cell morphology

The stenosis model allows the observation of the morphological response of the ECs to WSS gradients. Figure 7 shows the WSS patterns within the model as a function of position. A positive shear gradient is found in the proximal region of the stenosis, reaching a peak WSS just upstream of the apex. A negative WSS gradient is observed in the recirculation zone of the stenosis, however, downstream of the flow reattachment point, laminar Hagen-Poiseuille flow resumes.

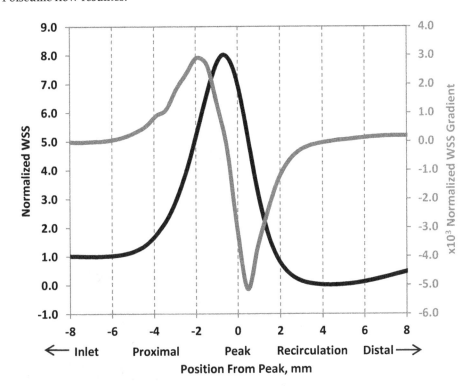

Fig. 7. CFD simulation of the WSS and WSS gradient Profile along the bottom central axis of the stenosis model (– Normalized WSS Magnitude; – Normalized WSS gradient).

It was found that after sufficient time had passed, morphological trends formed throughout the model. The inlet and outlet of the model showed similar shape indices as the straight tube controls, making them acceptable internal controls. These uniform internal controls can be compared to the other regions of the models. The deceleration, or recirculation zone (depending on the flow patterns, governed by the Re), showed the highest shape index, perhaps indicative of the most inflammatory response. The acceleration, or proximal zone, showed a slightly elevated shape index compared to the control regions, though this value was still statistically lower than that found in the deceleration zone. It was expected that the increased shear in this zone would result in more elongated ECs, however the results suggest that WSS gradients can have a more drastic effect on endothelium health than WSS magnitude alone.

It can be concluded that the deceleration zone could potentially present an inflamed endothelium and therefore one would predict to see an increase in regional expression of proteins linked to inflammation in that region. In turn, this should also lead to the largest amount of neutrophil adhesion in the recirculation zone.

3.2.2 Regional inflammation and adhesion molecule expression

WSS magnitude and duration was investigated to determine the effect these factors have on inflammatory response. Straight, cylindrical models were perfused under 4.5, 9 and 18 dyne/cm² inlet WSS conditions for various time periods (up to 24 hours). VCAM-1 and ICAM-1 mRNA expression decreased with increasing WSS magnitude and time, indicative of an atheroprotective phenotype. An increase in WSS magnitude resulted in a decrease in E-selectin mRNA expression; however, E-selectin expression increased from 0-12 hours, and sharply fell by 24 hours.

Inflammation was considered in the stenosis model by observing regional endothelial CAM expression using immunostaining and confocal microscopy. Perfusions with neutrophils were run at 1.25 dyne/cm², revealing an increase in CAM expression at the peak of the stenosis, Figure 8 (Rouleau et al., 2010a). Furthermore, perfusions were run at a higher inlet WSS of 6.25 dyne/cm², with no noticeable difference in regional CAM expression. These WSS values are consistent with the conditions used for neutrophil adhesion experiments.

TNF-α Stimulated 24 hrs – Static – Regional Analysis

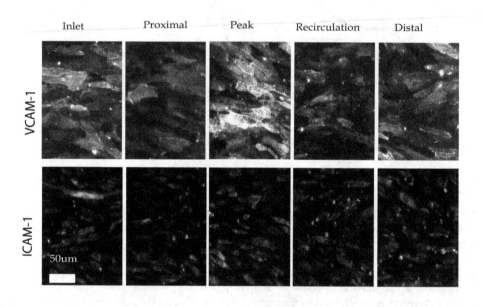

Fig. 8. Regional ICAM and VCAM expression. Copyright Springer, Annals of Biomedical Engineering, 38, 2010, pp. 2797, Neutrophil Adhesion on Endothelial Cells in a Novel Asymmetric Stenosis Model: Effect of Wall Shear Stress Gradients. Rouleau, L.; Copland, I; Tardif, J-C.; Mongrain, R. & Leask, R., Figure 6 with kind permission from Springer Science+Business Media B.V.

Using inlet WSS values of 4.5, 9 and 18 dyne/cm^2, more physiologically relevant hemodynamics were present. During these perfusions, ICAM-1, VCAM-1 and E-selectin levels were quantified and NF-κB translocation was observed to provide a robust picture of the regional inflammation around the stenosis. Similar to the experiments at 1.25 and 6.25 dyne/cm^2, there was an upregulation of ICAM-1 and VCAM-1 at the stenosis peak and in the proximal and recirculation zones. Similarly, E-selectin and NF-κB were also upregulated in these areas. It can then be concluded that there is a higher likelihood of increased neutrophil adhesion around the stenosis of the model.

3.2.3 Regional neutrophil adhesion

In perfusion experiments, it was shown that both WSS magnitude and perfusion time significantly affected the adhesion of a leukocyte cell line (NB4 cells). The trends showed that flow conditioned cells resulted in reduced adhesion of the NB4 cells for both the TNF-α stimulated and non-stimulated ECs. The two WSS magnitudes investigated, 1.25 dynes/cm^2 and 6.25 dynes/cm^2, resulted in a 3 fold and 15 fold decrease in adhesion, when compared to cells that were kept static prior to the adhesion experiments, respectively. Furthermore, experiments were run under even higher WSS conditions (12.5 dynes/cm^2) and it was found that very few cells were able to adhere to the ECs. This data demonstrates the influence of hemodynamic and attractive (ligand-receptor) forces acting on the neutrophils. The higher hemodynamic forces push the neutrophils off of the binding sites, overcoming the attractive forces which form during the adhesion of the neutrophils. For both the low (1.25 dynes/cm^2) and high (6.25 dynes/cm^2) shear stress conditions it was found that an increase in perfusion time from 1 to 6 hours resulted in an increase in adhesion, with a more noticeable increase occurring in non-stimulated ECs, potentially showing that at the shorter time point in TNF-α stimulated cells, a maximum *in vitro* adhesion is reached.

Regional neutrophil adhesion

The stenosis model presents a unique 3D environment which allowed for the investigation of the spatial differences in adhesion on and around a stenosis. Videos of the adhesion assays showed that there was a region of flow recirculation downstream of the stenosis. Immediately downstream of the separation point there was a distinct line of NB4 cell adhesion. It is postulated that this focal neutrophil adhesion was facilitated by low WSS and minimal fluid momentum caused by backflow in the recirculation zone.

For the rest of the analysis, the average adhesion was evaluated for each region of the model, Figure 9. TNF-α stimulation increased adhesion in all regions save for the stenosis peak. It was found that both the WSS magnitude and perfusion duration affected the incidence of adhesion. For example, it was found that at low inlet WSS (1.25 dyne/cm^2), ECs in the recirculation and distal regions showed a significant increase in adhesion from 1 to 6 hours.

It was found that in general, the recirculation zone tended to have the highest cell adhesion. It is hypothesized that the recirculation of NB4 cells results in a higher concentration of cells flowing along the endothelium (Rouleau et al., 2010a). Furthermore, the leukocytes have a lower momentum in the recirculation zone due to the decreased shear. This would allow for an increase in adhesion in this region. The endothelium in this location is also exposed to reduced WSS, leading to an increased inflammatory response.

The proximal and distal regions have lower incidence of adhesion than the recirculation region but more than at the stenosis peak. Interestingly, there seemed to be greater adhesion in the proximal region than the inlet of the model which may be due to a positive wall shear gradient. As the fluid reaches the stenosis, the projected surface area decreases resulting in an increase in WSS. Hinds et al. found a similar result in their studies using monocytes (Hinds et al., 2001). Comparing these two results, it can be seen that there is increased adhesion of leukocytes to ECs in the presence of complex wall shear stress gradients.

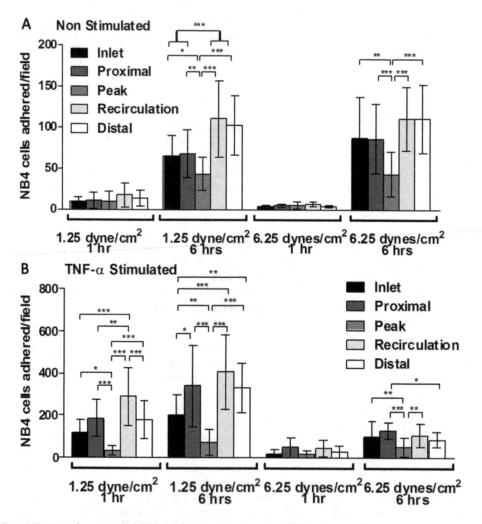

Fig. 9. Regional neutrophil (NB4) adhesion in the asymmetric tissue culture model. Copyright Springer, Annals of Biomedical Engineering, 38, 2010, pp. 2798, Neutrophil Adhesion on Endothelial Cells in a Novel Asymmetric Stenosis Model: Effect of Wall Shear Stress Gradients. Rouleau, L.; Copland, I; Tardif, J-C.; Mongrain, R. & Leask, R., Figure 7 (c) & (d) with kind permission from Springer Science+Business Media B.V.

In all instances, the stenosis peak had relatively low adhesion, which is a result of the high shear forces. By the peak of the stenosis, WSS values were appreciably larger than those found in the inlet (Figure 7). These hemodynamic forces exceed the adhesive force needed for the neutrophils to adhere. Extending static adhesion past 1 hour resulted in little or no additional adhesion. This was likely due to a lack of adhesion sites for the NB4s to bind to. Ultimately, under any conditions, there will be a point where the endothelium becomes saturated with bound NB4s and simply cannot facilitate further adhesion, *in vitro*. Although this leaves a certain limitation on the results of the aforementioned experiments, it also demonstrates the high levels of focal adhesion which can occur as a result of endothelial inflammation.

4. Conclusions

Atherosclerosis is an inflammatory disease. *In vitro* studies of the interaction of inflammatory cells with the endothelium have advanced our understanding of the role of inflammation in atherosclerosis development and progression. Our novel three dimensional dynamic cell culture model of a coronary stenosis has shown the importance of spatial gradients in wall shear stress in EC response and leukocyte attachment. Leukocyte attachment is increased in the proximal and distal regions of the stenosis. The increased attachment occurs in regions where the ECs have an inflamed phenotype. The results suggest that the hemodynamics created by the stenosis geometry create an inflammatory response of the endothelial cells that promotes leukocyte attachment. These results help to explain disease stability in established coronary stenoses.

5. References

Lloyd-Jones, D.; Adams, R.; Brown, T.; Carnethon, M.; Dai, S.; De, S.; Ferguson, T.; Ford, E.; Furie, K.; Gillespie, C.; Go, A.; Greenlund, K.; Haase, N.; Hailpern, S.; Ho, P.; Howard, V.; Kissela, B.; Kittner, S.; Lackland, D.; Lisabeth, L.; Marelli, A.; McDermott, M.; Meigs, J.; Mozaffarian, D.; Mussolino, M.; Nichol, G.; Roger, V.; Rosamond, W.; Sacco, R.; Sorlie, P.; Roger, V.; Thom, T.; Wasserthiel-Smoller, S.; Wong, N. & Wylie-Rosett, J. (2010). Heart disease and stroke statistics--2010 update: a report from the American Heart Association. *Circulation*, 121, 7, (February 2010), pp. (e46-e215)

Waller, B.; Orr, C.; Slack, J.; Pinkerton, C.; Van, T. & Peters, T. (1992). Anatomy, histology, and pathology of coronary arteries: a review relevant to new interventional and imaging techniques--Part I. *Clin Cardiol*, 15, 6, (June 1992), pp. (451-457)

Shah, P. (2006). Pathogenesis of Atherosclerosis, In: *Essential Cardiology*, C. Rosendorff, pp. (409-418), Humana Press, Retrieved from <http://dx.doi.org/10.1007/978-1-59259-918-9_22>

Libby, P. (2002). Inflammation in atherosclerosis. *Nature*, 420, 6917, (December 2002), pp. (868-874)

Van der Heiden, K.; Cuhlmann, S.; Luong, l.; Zakkar, M. & Evans, P. (2010). Role of nuclear factor kappaB in cardiovascular health and disease. *Clin Sci (Lond)*, 118, 10, (May 2010), pp. (593-605)

Schwartz, C.; Valente, A.; Sprague, E.; Kelley, J. & Nerem, R. (1991). The pathogenesis of atherosclerosis: an overview. *Clin Cardiol*, 14, 2 Suppl 1, (February 1991), pp. (I1-16)

Libby, P.; Ridker, P. & Maseri, A. (2002). Inflammation and Atherosclerosis. *Circulation*, 105, 9, (March 2002), pp. (1135-1143)

Ross, R. (1993). The pathogenesis of atherosclerosis: a perspective for the 1990s. *Nature*, 362, 6423, (April 1993), pp. (801-809)

Ross, R. (1999). Atherosclerosis--an inflammatory disease. *N Engl J Med*, 340, 2, (January 1999), pp. (115-126)

Ross, R.; Glomset, J. & Harker, L. (1977). Response to injury and atherogenesis. *Am J Pathol*, 86, 3, (March 1977), pp. (675-684)

Kelly, M.; Hwang, J. & Kubes, P. (2007). Modulating leukocyte recruitment in inflammation. *J Allergy Clin Immunol*, 120, 1, (2007), pp. (3-10)

Miyasaka, M.; Kawashima, H.; Korenaga, R. & Ando, J. (1997). Involvement of selectins in atherogenesis: a primary or secondary event? *Ann N Y Acad Sci*, 811, (April 1997), pp. (25-34)

Kubes, P. & Kerfoot, S. (2001). Leukocyte recruitment in the microcirculation: the rolling paradigm revisited. *News Physiol Sci*, 16, (April 2001), pp. (76-80)

Rao, R.; Yang, L.; Garcia-Cardena, G. & Luscinskas, F. (2007). Endothelial-dependent mechanisms of leukocyte recruitment to the vascular wall. *Circ Res*, 101, 3, (August 2007), pp. (234-247)

Iiyama, K.; Hajra, L.; Iiyama, M.; Li, H.; DiChiara, M.; Medoff, B. & Cybulsky, M. (1999). Patterns of vascular cell adhesion molecule-1 and intercellular adhesion molecule-1 expression in rabbit and mouse atherosclerotic lesions and at sites predisposed to lesion formation. *Circ Res*, 85, 2, (July 1999), pp. (199-207)

Schenkel, A.; Mamdouh, Z. & Muller, W. (2004). Locomotion of monocytes on endothelium is a critical step during extravasation. *Nat Immunol*, 5, 4, (April 2004), pp. (393-400)

Glagov, S.; Zarins, C.; Giddens, D. & Ku, D. (1988). Hemodynamics and atherosclerosis. Insights and perspectives gained from studies of human arteries. *Arch Pathol Lab Med*, 112, 10, (October 1988), pp. (1018-1031)

Nichols, W. & O'Rourke, M. (1990). McDonald's Blood Flow in Arteries; theoretical, experimental and clinical principles. Third Edition, (1990), pp. (1-456), Lea & Febiger, Philadelphia

Ku D. (1997). Blood flow in arteries. *Annual Review of Fluid Mechanics*, 29, 1, (1997), pp. (399-434)

Malek, A. M.; Alper, S. L. & Izumo, S. (1999). Hemodynamic shear stress and its role in atherosclerosis. *JAMA*, 282, 21, (December 1999), pp. (2035-2042)

Myers, J.; Moore, J.; Ojha, M.; Johnston, K. & Ethier, C. (2001). Factors influencing blood flow patterns in the human right coronary artery. *Ann Biomed Eng*, 29, 2, (February 2001), pp. (109-120)

Firrell, J. & Lipowsky, H. (1989). Leukocyte margination and deformation in mesenteric venules of rat. *Am J Physiol*, 256, 6 Pt 2, (June 1989), pp. (H1667-H1674)

Lawrence, M.; Kansas, G.; Kunkel, E. & Ley, K. (1997). Threshold levels of fluid shear promote leukocyte adhesion through selectins (CD62L,P,E). *J Cell Biol*, 136, 3, (February 1997), pp. (717-727)

Evans, E.; Leung, A.; Heinrich, V. & Zhu, C. (2004). Mechanical switching and coupling between two dissociation pathways in a P-selectin adhesion bond. *Proc Natl Acad Sci U S A*, 101, 31, (August 2004), pp. (11281-11286)

Cozens-Roberts, C.; Quinn, J. & Lauffenberger, D. (1990). Receptor-mediated adhesion phenomena. Model studies with the Radical-Flow Detachment Assay. *Biophys J*, 58, 1, (July 1990), pp. (107-125)

Tees, D. & Goetz, D. (2003). Leukocyte adhesion: an exquisite balance of hydrodynamic and molecular forces. *News Physiol Sci*, 18, (October 2003), pp. (186-190)

Marshall, B.; Long, M.; Piper, J.; Yago, T.; McEver, R. & Zhu, C. (2003). Direct observation of catch bonds involving cell-adhesion molecules. *Nature*, 423, 6936, (May 2003), pp. (190-193)

Sarangapani, K.; Yago, T.; Klopocki, A.; Lawrence, M.; Fieger, C.; Rosen, S.; McEver, R. & Zhu, C. (2004). Low force decelerates L-selectin dissociation from P-selectin glycoprotein ligand-1 and endoglycan. *Journal of Biological Chemistry*, 279, 3, (January 2004), pp. (2291-2298)

Lawrence, M.; McIntire, L. & Eskin, S. (1987). Effect of flow on polymorphonuclear leukocyte/endothelial cell adhesion. *Blood*, 70, 5, (November 1987), pp. (1284-1290)

Bacabac, R.; Smit, T.; Cowin, S.; Van Loon, J.; Nieuwstadt, F.; Heethaar, R. & Klein-Nulend, J. (2005). Dynamic shear stress in parallel-plate flow chambers. *J Biomech*, 38, 1, (January 2005), pp. (159-167)

Forrester, J. & Lackie, J. (1984). Adhesion of neutrophil leucocytes under conditions of flow. *J Cell Sci*, 70, (August 1984), pp. (93-110)

Finger, E.; Puri, K.; Alon, R.; Lawrence, M.; von Andrian, U. & Springer, T. (1996). Adhesion through L-selectin requires a threshold hydrodynamic shear. *Nature*, 379, 6562, (January 1996), pp. (266-269)

Alon, R.; Hammer, D. & Springer, T. (1995). Lifetime of the P-selectin-carbohydrate bond and its response to tensile force in hydrodynamic flow. *Nature*, 374, 6522, (April 1995), pp. (539-542)

Sheikh, S.; Rahman, M.; Gale, Z.; Luu, N.; Stone, P.; Matharu, N.; Rainger, G. & Nash, G. (2005). Differing mechanisms of leukocyte recruitment and sensitivity to conditioning by shear stress for endothelial cells treated with tumour necrosis factor-alpha or interleukin-1beta. *Br J Pharmacol*, 145, 8, (August 2005), pp. (1052-1061)

Sheikh, S.; Rainger, G.; Gale, Z.; Rahman, M. & Nash, G. (2003). Exposure to fluid shear stress modulates the ability of endothelial cells to recruit neutrophils in response to tumor necrosis factor-alpha: a basis for local variations in vascular sensitivity to inflammation. *Blood*, 102, 8, (October 2003), pp. (2828-2834)

Alcaide, P.; Auerbach, S. & Luscinskas, F. (2009). Neutrophil Recruitment under Shear Flow: It's All about Endothelial Cell Rings and Gaps. *Microcirculation*, 16, 1, (2009), pp. (43-57)

Chen, C. N.; Chang, S.; Lee, P.; Chang, K.; Chen, L.; Usami, S.; Chien, S. & Chiu, J. (2006). Neutrophils, lymphocytes, and monocytes exhibit diverse behaviors in transendothelial and subendothelial migrations under coculture with smooth muscle cells in disturbed flow. *Blood*, 107, 5, (March 2006), pp. (1933-1942)

Burns, M. & DePaola, N. (2005). Flow-conditioned HUVECs support clustered leukocyte adhesion by coexpressing ICAM-1 and E-selectin. *AJP - Heart and Circulatory Physiology*, 288, 1, (January 2005), pp. (H194-H204)

Shankaran, H. & Neelamegham, S. (2001). Effect of secondary flow on biological experiments in the cone-plate viscometer: methods for estimating collision frequency, wall shear stress and inter-particle interactions in non-linear flow. *Biorheology*, 38, 4, (2001), pp. (275-304)

Walpola, P.; Gotlieb, A.; Cybulsky, M. & Langille, B. (1995). Expression of ICAM-1 and VCAM-1 and monocyte adherence in arteries exposed to altered shear stress. *Arteriosclerosis, Thrombosis, and Vascular Biology*, 15, 1, (January 1995), pp. (2-10)

Walpola, P.; Gotlieb, A. & Langille, B. (1993). Monocyte adhesion and changes in endothelial cell number, morphology, and F-actin distribution elicited by low shear stress in vivo. *Am J Pathol*, 142, 5, (May 1993), pp. (1392-1400)

Hinds, M.; Park, Y.; Jones, S.; Giddens, D. & Alevriadou, B. (2001). Local hemodynamics affect monocytic cell adhesion to a three-dimensional flow model coated with E-selectin. *J Biomech*, 34, 1, (January 2001), pp. (95-103)

Rouleau, L.; Copland, I.; Tardif, J.; Mongrain, R. & Leask, R. (2010). Neutrophil adhesion on endothelial cells in a novel asymmetric stenosis model: effect of wall shear stress gradients. *Ann Biomed Eng*, 38, 9, (September 2010), pp. (2791-2804)

Rouleau, L.; Farcas, M.; Tardif, J.; Mongrain, R. & Leask, R. (2010). Endothelial cell morphologic response to asymmetric stenosis hemodynamics: effects of spatial wall shear stress gradients. *J Biomech Eng*, 132, 8, (August 2010), pp. (081013-

Brunette, J.; Mongrain, R.; Laurier, J.; Galaz, R. & Tardif, J. (2008). 3D flow study in a mildly stenotic coronary artery phantom using a whole volume PIV method. *Medical Engineering & Physics*, 30, 9, (November 2008), pp. (1193-1200)

Wexler, L.; Brundage, B.; Crouse, J.; Detrano, R.; Fuster, V.; Maddahi, J.; Rumberger, J.; Stanford, W.; White, R. & Taubert, K. (1996). Coronary artery calcification: pathophysiology, epidemiology, imaging methods, and clinical implications. A statement for health professionals from the American Heart Association. Writing Group. *Circulation*, 94, 5, (September 1996), pp. (1175-1192)

Rouleau, L.; Rossi, J. & Leask, R. (2010). Concentration and time effects of dextran exposure on endothelial cell viability, attachment, and inflammatory marker expression in vitro. *Ann Biomed Eng*, 38, 4, (April 2010), pp. (1451-1462)

Choudhury, N.; Bouchot, O.; Rouleau, L.; Tremblay, D.; Cartier, R.; Butany, J.; Mongrain, R. & Leask, R. (2009). Local mechanical and structural properties of healthy and diseased human ascending aorta tissue. *Cardiovasc Pathol*, 18, 2, (March 2009), pp. (83-91)

Tremblay, D.; Cartier, R.; Mongrain, R. & Leask, R. (2010). Regional dependency of the vascular smooth muscle cell contribution to the mechanical properties of the pig ascending aortic tissue. *J Biomech*, 43, 12, (August 2010), pp. (2448-2451)

Ethier C.; Prakash, S.; Steinman, D.; Leask, R.; Couch, G. & Ojha, M. (2000). Steady Flow Separation Patterns in a 45 Degree Junction. *J Fluid Mechanics*, 411, (2000), pp. (1-38)

Mahinpey, N.; Leask, R.; Ojha, M.; Johnston, K. & Trass, O. (2004). Experimental study on local mass transfer in a simplified bifurcation model: potential role in atherosclerosis. *Ann Biomed Eng*, 32, 11, (November 2004), pp. (1504-1518)

Couch G.; Johnston, K. & Ojha, M. (1996). Full-field flow visualization and velocity measurement with a photochromic grid method. *Meas Sci Technol*, 7, (1996), pp. (1238-1246)

Park.H.; Moore J.; Trass O. & Ojha, M. (1999). Laser photochromic velocimetry estimation of the vorticity and pressure field-two dimenstional flow in a curved vessel. *Experiments in Fluids*, 26, (1999), pp. (55-62)

Dartsch, P. & Betz, E. (1989). Response of cultured endothelial cells to mechanical stimulation. *Basic Res Cardiol*, 84, 3, (May 1989), pp. (268-281)

Nerem, R.; Levesque, M. & Cornhill, J. (1981). Vascular endothelial morphology as an indicator of the pattern of blood flow. *J Biomech Eng*, 103, 3, (August 1981), pp. (172-176)

Levesque, M. & Nerem, R. (1985). The elongation and orientation of cultured endothelial cells in response to shear stress. *J Biomech Eng*, 107, 4, (November 1985), pp. (341-347)

Levesque, M.; Liepsch, D.; Moravec, S. & Nerem, R. (1986). Correlation of endothelial cell shape and wall shear stress in a stenosed dog aorta. *Arteriosclerosis*, 6, 2, (March 1986), pp. (220-229)

Helmlinger, G.; Geiger, R.; Schreck, S. & Nerem, R. (1991). Effects of pulsatile flow on cultured vascular endothelial cell morphology. *J Biomech Eng*, 113, 2, (May 1991), pp. (123-131)

Nerem, R. (1993). Hemodynamics and the vascular endothelium. *J Biomech Eng*, 115, 4B, (November 1993), pp. (510-514)

9

Pain in Osteoarthritis: Emerging Techniques and Technologies for Its Treatment

Kingsley Enohumah
The Rotunda Hospital, Dublin
Ireland

1. Introduction

Osteoarthritis (OA) is described as a condition characterised by use-related joint pain experienced on most days in any given month, for which no other cause is apparent. OA is the commonest disease affecting synovial joints and affects more than 40% of the population over 65 years. It affects primary the knee joints however hip, ankle, shoulder and small joints of hand and feet may be involved.

Previously, OA was considered a wear and tear, degenerative disease that must be accepted as an inevitable consequence of trauma and ageing. With advances in research and understanding of the mechanism of OA progression it is now known as a disease of the synovial joint affecting subchondral bone, synovium, meniscus, ligaments and supporting structures around the joints, including the cartilage.

The pathological changes seen in OA are characterised by focal areas of loss of articular cartilage within the synovial joints, associated with hypertrophy of the bone (osteophytes and subchondral sclerosis) and thickening of the capsule. OA is a chronic, degenerative disease associated with joint pain and loss of function. The primary problem in OA is the damage to the articular cartilage, which triggers a series of other events that culminate in pain and loss/limitation of function in the affected joint.

Undoubtedly, pain, which is the most prominent and disabling presentation of OA, is an increasingly important public health problem especially within an increasing aging population.

2. Epidermiology

OA occurs worldwide with higher prevalence in developed societies. Its twice as common in women as in men with a significant familial tendency.

The prevalence of OA increases with age in a progressive manner with 80% radiographic changes in people by the age of 65 years. However only about 25-30% are symptomatic. Primary OA is uncommon before the age of 50 years.

World Health Organisation reports that knee OA is ranked fourth most important global cause of disability in women and the eighth most important in men. Annual arthroplasty rate in over the age of 65 in Europeans vary from country to country but are of the order of

0.5–0.7 per 1000. The annual costs attributable to knee OA are immense. There is therefore a burden on health from both morbidity and cost.

OA is a complex disorder with multiple risk factors.

2.1 Risk factors for OA

1. Age > 50 years
2. Crystals in joint fluid or cartilage
3. High bone mineral density
4. History of immobilisation
5. Injury to the joint
6. Joint hypermobility or instability
7. Obesity (weight-bearing joints)
8. Peripheral neuropathy
9. Prolonged occupational or sports stress

3 Anatomy of a joint

A brief review of the basic anatomy of a typical synovial joint is presented here to help understand the mechanisms involved in OA-induced damages of the involved joint that culminate in pain and other symptoms of OA.

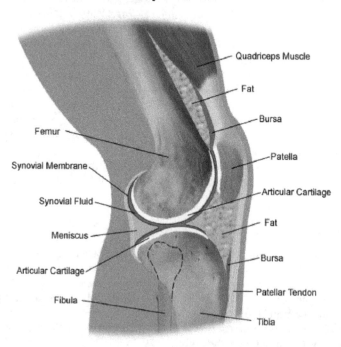

Fig. 1. A typical synovial joint

A joint is where two bones meet. Articular cartilage covers the bone ends which are lubricated by synovial fluid. Seventy to eighty per cent of the cartilage is made up of water and a type II collagen with proteoglycans and glycosaminoglycans produced by chrondrocytes. The collagen fibres in the cartilage offer tensile strength to the cartilage because of its architectural makeup. The cartilage, however, contains no intrinsic blood vessels. It receives its nutrition from the synovial fluid. The synovial fluid, which is secreted by the synovial membrane lining the inner surface of the joint, facilitates not only movement but also provides nutrients, phagocytosis and other immunologic functions within the joint. The integrity of a joint is therefore dependent upon its architecture, the cartilage, bone and the supporting structures enclosing the joint. OA in simple terms is a result of alterations in the aforementioned architectural structures within the joint with resultant pain, loss of function and instability in the involved joint. Figure 1 shows the diagram of a typical synovial joint.

4. Sources of nociception in a joint

Pain is defined as "an unpleasant sensory and emotional experience associated with actual or potential tissue damage, or described in terms of such damage". Pain, as generally acknowledged, is mainly a signal that the body has been injured.

The term "nociception" was coined by the Nobel Laureate Sherrington to designate a physiological sensory phenomenon. "Nociception" is derived from "nocere", the Latin word for "to hurt". Nociceptors are peripheral sensory organs that are activated when nociceptive stimuli cause tissue damage. These nociceptors are unspecialised, naked nerve endings found close to small blood vessels and mast cells. The functional nociceptive unit is therefore made up of the structural triad of capillary, nociceptor and mast cell. This is the unit that is sensitive to tissue damage. There are also a rich supply of myelinated and unmyelinated fibres innervating the joint capsule, ligaments subchondral bone, periosteum and menisci.

In the anatomy of the joint described above, the cartilage does not contain blood vessels but derives its nutrients from the synovium. The subchondral bone, periosteum, synovium, ligaments, and the joint capsule contain nerve endings that could be the source of nociceptive stimuli in OA. Irritation of the periostal as a result of remodelling, denuded bone, compression of soft tissue by osteophytes, microfractures of the subchondral bone, effusion and spasm of surrounding muscles has been shown to contribute to the pain that may be felt by patients with OA. So in effect the bone in the periosteum and bone marrow is richly innervated with nociceptive fibres and represents a potential source of nociceptive pain in patients with OA.

5. Pathology and pathogenesis

OA is a heterogenous spectrum of clinical condition affecting mostly joints. No one mechanism explains the various processes seen in the joint of OA. Factors including inflammation, genetic, injury or trauma and joint mechanics have all been implicated in the pathophysiology of OA. Each joint response is a balance of the anabolic and catabolic factors acting in combination with both the extrinsic and intrisic factors.

Summary of the mechanisms suggested for the pathogenesis of OA are:

* *Matrix loss:* Metalloproteinases (MMPs) such as stromelysin and collagenase which are secreted by the chondrocytes catalyses the degradation of both collagen and proteoglycans resulting in matrix loss

- *Role of inflammatory mediators:* Mediators such as TNF-α and IL-1 stimulate MMPs secretion and this inhibit collagen production.
- *Tissue inhibitors of MMPs:* Tissue inhibitors of MMPs regulate the MMPs. Therefore any disturbance of this regulatory mechanism may lead to increased cartilage degradation and may contribute to the development of OA.
- *Growth factors deficiency:* Growth factors such as insulin-like growth factor and transforming growth factor enhance collagen synthesis and so when these factors are deficient matrix repair is impaired.
- Genetic susceptibility

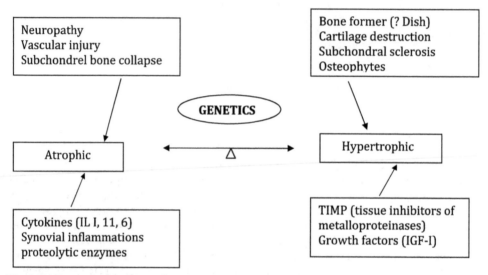

Fig. 2. Different factors that influence OA process.

Stage I	There is proteolytic breakdown of cartilage matrix
Stage II	There is fibrillation and erosion of cartilage surface, accompanied by the release of breakdown products into the synovial fluid
Stage III	Synovial inflammation begins when synovial cells ingest a breakdown product through phagocytosisand produce proteases and proinflammatory cytokines

Table 1. Stages of OA (Martel-Pelletier, 2004)

5.1 Classification of OA

Primary OA	Has no known cause. Common. Related to aging and hereditary. May be localised or generalised. Commonly affects the distal interphalangeal joints of the hands, hip and the knee. The cervical and lumbar spine may be affected.
Secondary OA	Causes include articular injury, obesity, Paget's disease, or inflammatory arthritis and aging process. May be localised or generalised. May affect any joint and can occur at any age.

5.2 Clinical features

Pain and functional restriction are the main symptoms in OA. The pain is characteristically made worse by movement and relieved by rest.

Signs	Symptoms
Joint tenderness	Joint pain
Crepitus on movement	Joint gelling (stiffening and pain after mobility)
Limitation of range of movement	Joint instability
Joint instability	Loss of function
Joint effusion and variable levels of inflammation	
Bone swelling	
Wasting of muscles	

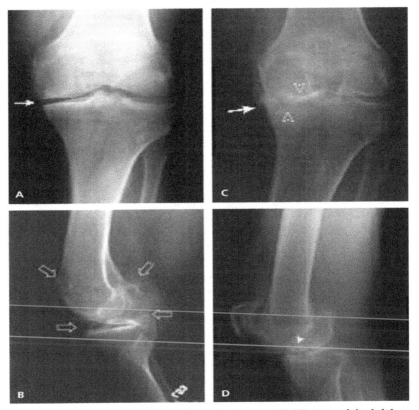

Fig. 3. Radiological changes in Osteoarthritis of the knees. (A) AP view of the left knee shows medial joint space narrowing (arrow). (B) Lateral view shows sclerosis with marked osteophyte formation (arrows). (C) Medial joint space narrowing (white arrow) causing a varus deformity of the knee and collapse of the joint space with destruction of the medial cartilage and the subchondral cortex (open arrow heads). (D) Subchondral cysts are noted (solid arrow head).

5.3 Radiological changes in OA
- Joint space narrowing
- Osteophytes
- Bony cysts
- Subchondral sclerosis

5.4 Mechanism of pain in OA

Summary of process of pain perception

- A noxious stimulus causes stimulation of nociceptors (pain receptors) in the receptor organ (e.g. joint).
- This firing of primary afferent fibres at the site of tissue injury causes axonal release of substance P (SP). This stimulation leads to activation of cells in the dorsal horn of the spinal cord and transmission of the nerve impulse to the midbrain and cortex. Thus, impulses travelling along first order neuron synapse on second-order neuron in the dorsal horn of the spinal cord. The axon crosses to the contralateral side and ascends to synapse on the third-order neurons. The third-order neurons send fibres to the cerebral cortex where conscious perception of the sensation occurs.
- Transmission of sensory information is modulated (inhibited or potentiated) throughout the nervous system by neurons from the midbrain and spinal cord that release endogenous opioids, catecholamines and other neurotransmitters.
- Peripheral nociceptor sensitisation, which is the transmission of impulses at subnormal threshold, occurs following the release of chemical mediators such as prostagladins and leukotrienes at the site of injury or damage. Continued stimulation by peripheral nociceptors then leads to sensitisation of neurons in the spinal cord. This is known as central sensitisation.

Tissue injury results in the release of inflammatory mediators such as serotonin, bradykinin, calcitonin gene-related peptide (CGRP) and SP, which lead to nociceptor nerve fibre sensitisation in peripheral tissue. These damaged fibres release inflammatory agents causing a spread of increased sensitivity around the area of tissue damage. This is called primary hyperalgesia. The repeated depolarisation of primary afferent fibres leads to a continuous release of neurotransmitters onto the secondary neurons in the spinal cord, resulting in central sensitisation and secondary hyperalgesia. Peripheral pain sensitisation is a feature of osteoarthritis in the joint.

In addition to peripheral pain sensitisation pain in OA, could also be due to local and central sensitisation of pain, pathways resulting in normal stimuli becoming painful with inflammation being an important feature in the process of OA.

Most of the substances involved in inflammation such as proinflammatory cytokines and bradykinins interact with the nociceptive fibres present within the joint and induce hyperalgesia and allodynia seen in patients with chronic inflammatory joint disease like OA. These mechanisms acting in concert could participate in the progression of hyperalgesia to chronicity.

5.5 Progression of OA to chronicity
Chronic pain (CP) is pain that persists for a month beyond the usual course of an acute disease or a reasonable time period for an injury to heal.

CP differs from the acute process not only in the duration of its course but also the different receptors involved in the mechanisms of action for acute pain and CP.

Those most involved in the acute process are a-amino-3-hydroxy- 5-methyl-isoxazole-4-propionic acid (AMPA) receptors, while those of primary importance in the sensation of CP are N-methyl- D-aspartate (NMDA) receptors. Activation of NMDA receptors causes the release of peptide neurotransmitter SP, which amplifies the pain by causing the spinal neurons carrying the pain to be easily stimulated.

Elevated levels of SP in spinal fluids have been documented in patients with OA and fibromyalgia. The progression of nociception from an acute to a chronic process has yet to be fully understood. However, recent evidence from animal experiments as well as human research suggests that peripheral mechanisms in acute pain and long-term potentiation (LTP) of neuronal sensitivity to nociceptive inputs in the dorsal horn of the spinal cord may underline the transition from acute to a chronic process.

LTP in spinal nociceptive systems has been suggested as one of the mechanisms underpinning the transition of acute pain to CP. It seems possible that LTP may underlie some forms of afferent induced hyperalgesia and that simultaneous activation of NMDA; SP neurokinin-I (NK-I) and glutamate receptors are required for the induction of spinal LTP. Therefore, it is likely that the conditioning stimuli that induce synaptic LTP in the superficial spinal dorsal horn are similar to those that trigger hyperalgesia. LTP is likely to occur in both the sensory and the affective pain pathways. Additionally, spinal LTP and injury-induced hyperalgesia share signal transduction pathways, which make use-dependent LTP an attractive model of injury-induced central sensitisation and hyperalgesia.

Summary of progression to chronic pain state

1. Rapid, intense stimulation of CA1 neurons in the hippocampus depolarizes them.
2. Binding of Glu and D-serine to their NMDA receptors opens them.
3. Ca^{2+} ions flow into the cell through the NMDA receptors and bind to calmodulin.
4. This activates calcium-calmodulin-dependent kinase II (CaMKII).
   - CaMKII phosphorylates AMPA receptors making them more permeable to the inflow of Na^+ ions and thus increasing the sensitivity of the cell to depolarization.
   - In time CaMKII also increases the number of AMPA receptors at the synapse.
5. Increased gene expression (i.e., protein synthesis — perhaps of AMPA receptors) also occurs during the development of LTP.
6. Enlargement of the synaptic connections and perhaps the formation of additional synapses occur during the formation of LTP.

6. Treatment: emerging techniques and technologies

The modes of treatment for OA have always focus on decreasing pain and improving function ranging from information, education, physical therapy and aids, through analgesics, non-steroidal anti-inflammatory drugs and joint injections, and to surgery in which all or part of the joint is replaced with plastic, metal or ceramic implants.

OA is complex in genetics, pathogenesis, monitoring and treatment however, the principal goals of management are:

- Education of the patient about OA
- Pain relief
- Achieving and maintaining optimal joint and limb function
- Reducing adverse factors to beneficially modify the OA process and its outcome.

Despite huge laboratory and clinical research, there are no proven diseases modifying therapies for OA. However, emerging orthopaedic surgical procedures may help to alleviate the attendant pain and functional loss resulting from joint damage in OA.

Some of the surgical approaches in the management of OA include

- Arthroscopic approach
- Osteotomies
- Total joint replacements and arthrodesis
- Tissue engineering and biologic therapies
    - Autologous Chondrocyte Implantation (ACI)
    - Meniscal Transplantation (MT)

6.1 Arthroscopic procedures for OA

Arthroscopic surgery is a routine surgical procedure for joint debridement and lavage in the management of OA since the 1980s. The advent of this technique has permitted less invasive access to joints and the opportunity to intervene earlier in the course of joint destruction, potentially to delay and/or prevent a predictably progressive degenerative pathway. However, in recent times the only indication where this technique is thought to be of benefit is in the management of OA with a superimposed structural lesion such as a meniscal tear in which arthroscopic partial meniscectomy (APM) is performed simultaneously.

There is a strong research and clinical evidence that patients with symptoms attributable to knee OA per se, and not meniscal tear, do not improve following arthroscopic lavage and debridement. Whether APM is useful in patients with symptomatic meniscal tear and concomitant OA is unclear at this stage. This is an area of investigation at the moment.

6.2 Osteotomies

Osteotomies are performed to restore a more anatomic biomechanical environment and prevent or delay the onset of OA or slow its progression.

In symptomatic patients with OA, osteotomy is performed to realign joints with the aims of relieving pain and delaying the onset or progression of OA. Osteotomy and joint preserving surgical procedures should be considered in young adults with symptomatic OA, especially in the presence of dysplasia or varus/valgus deformity.

6.2.1 Indications

- As an adjunct in younger patients with predominantly unicompartmental OA
- Age less than 60 years
- 10 to 15 degrees of varus deformity on weight bearing radiographs
- Preoperative motion arc of at least 90 degrees
- Flexion contracture less than 15 degrees
- Ability and motivation to effectively and safely perform rehabilitation

6.2.2 Contraindications to osteotomy

- Lateral compartment loss of joint space
- Lateral tibial subluxation greater than 1 centimeter
- Medial bone loss greater than 2 to 3 millimeters
- Ligamentous instability
- Inflammatory arthritis

6.2.3 Surgical types and methods

Surgical Types	Methods
Medial compartment knee OA and varus deformity	High tibial osteotomy is performed either by removing a wedge of bone from the lateral proximal tibia or more commonly by opening wedge space in the a medial proximal tibia.
• Advantages	Permits the knee to adapt a more valgus alignment. Transfers load from the damaged medial compartment to the more normal cartilage of the lateral compartment.
Lateral compartment knee OA and varus deformity	This is a distal femoral osteotomy in which a wedge of bone is removed from the medial distal femur or a wedge is opened in the lateral aspect.
• Advantage	Shifts the load to the healthier medial compartment.

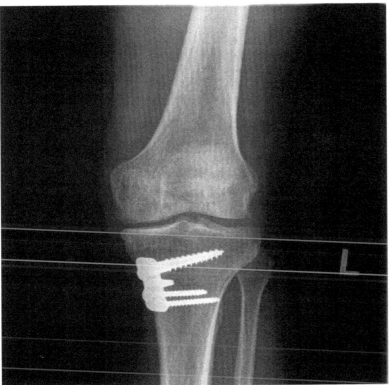

Fig. 4. AP radiograph of medial opening wedge high tibial osteotomy performed for medial compartment osteoarthritis.

6.3 Total joint replacements and arthrodesis

Total joint replacement (TJR) has to be considered in patients with radiographic evidence of hip/ knee OA who have refractory pain and disability. Principally, OA occurs less commonly at ankle, elbow, and wrist and thus total joint replacements are less frequent at these sites than at the hip or knee. In the last few years, interest in total knee arthroplasty has resulted in a proliferation of prosthetic designs, and many different types are now aaialble.

The indications for TJR have evolved and are expanding. Currently TJR are offered to patients earlier in the course of the disease as the risks of complications associated with TJR have reduced dramatically.

The prostheses available are:

1. *Condvlar replacements:* The joint surfaces alone are replaced. Ligaments then are needed to provide stability.
2. *Hinge-type prostheses:* In this type the ligaments are sacrificed and stability is provided by the design of the prosthesis itself.

The selection of a suitable prosthesis is dependent on the type and the indications

Types of prostheses

1. Unicondylar

This is an anatomically designed replacement for either the medial or the lateral femoral tibial articulation. It is designed to allow 120 degrees of flexion. The *unicondvlar* prosthesis is used only for cornpartmental OA.

2. Duocondylar

The femoral component of the duocondylar prosthesis is similar in shape to that of the unicondylar model except that there is no anterior flange and instead the halves are connected by an anterior cross bar which is countersunk during insertion. Because of its anatomical shape, it is most suitable when deformity, instability, and flexion contracture are not too severe.

3. Geometric

The prosthesis is non-anatomical in that the curvature of the femoral component is of constant radius. The plastic tibial component is in one piece, with two halves connected by an anterior bar. The prosthesis is designed to allow a 90- degree arc of motion. The cruciate ligaments are preserved. Two sizes are available.

4. Guepar

The Guepar is a Vitallium hinge prosthesis (improved over the Young model) which is fully constrained, providing motion in a fixed axis without rotation. Guepar prosthesis was used in knees with extrerne deformity or instability due to rheumatoid arthritis and OA.

Innovation continues to characterize the TJR field. This clinical dilemma has stimulated a search for biomaterials that produce less wear debris and, in turn, cause less osteolysis, attendant bone loss, and implant failure. This is the rationale for several developments, including highly cross-linked polyethylene and ceramic- on-ceramic and metal-on-metal bearing surfaces.

6.4 Surgical and biologic procedures

Advances in tissue engineering and biologic therapy have led to a few limited successes. Perhaps the most notable is autologous chondrocyte implantation (ACI).

Indications

1. Age <50 years
2. Isolated cartilage defects typically greater than 3 cm^2 in size

This procedure attempts to repair a symptomatic cartilage defect (Figure 3A) through implantation of chondrocytes grown ex vivo from a small cartilage biopsy sample obtained from the patient in a staging arthroscopy. After debridement of any degenerated tissue in the defect, a patch material, either periosteum from the patient or a synthetic collagen membrane, is sutured over the defect to create a watertight chamber into which the chondrocyte suspension is injected (Figure 3B). The chondrocytes attach to the subchondral bone and produce cartilage matrix, eventually filling the defect with hyaline-like cartilage.

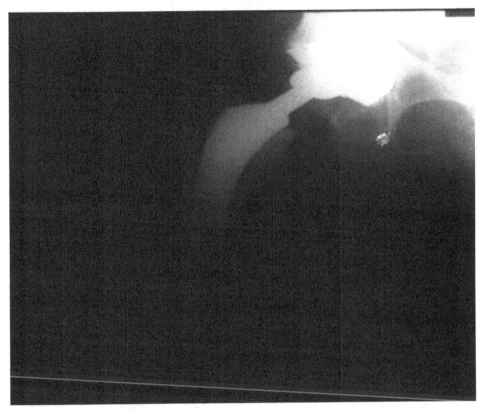

Fig. 5. Total hip replacement

6.5 Conclusion

It is clear from the foregoing that any simple unitary concept about the link between joint damage and symptoms in OA is untenable. We are faced with a complex interaction between local events in the joint, pain sensitisation, the cortical experience of pain, and what people are doing in their everyday lives.

In the absence of effective disease-modifying therapy, many patients with OA progress to advanced joint destruction. Therefore, surgery plays an important role in the management of OA. Advances in biomaterials and tissue engineering will continue to create exciting new opportunities to integrate surgical approaches in OA care.

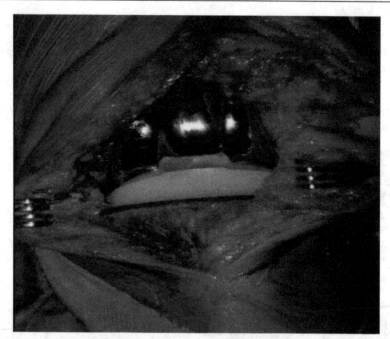

Fig. 6. Total knee replacement

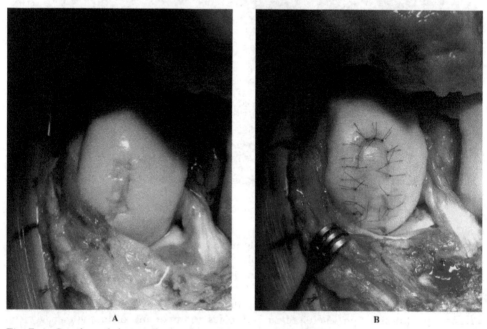

Fig. 7. **A**, Cartilage defect on femoral condyle.
B, Cartilage defect treated with autologous cartilage implantation

7. References

[1] Katz JN, Earp BE and Gomoll AH. Surgical management of Osteoarthritis. Arthritis Care & Research Vol. 62, No. 9, 2010, pp 1220–1228

[2] Haq I, Murphy E and Dacre J. Osteoarthritis. Postgrad Med J 2003; 79:377-383

[3] Kirkley A, Birmingham TB, Litchfield RB, Giffin JR, Willits KR, Wong CJ, et al. A randomized trial of arthroscopic surgery for osteoarthritis of the knee. N Engl J Med 2008; 359:1097-107.

[4] Felson DT, Lawrence RC, Dieppe PA, Hirsch R, Helmick CG, Jordan JM, et al. Osteoarthritis: new insights. Part 1: the disease and its risk factors. Ann Intern Med 2000; 133:635-46.

[5] W-Dahl A, Robertsson O, Lidgren L. Surgery for knee osteoarthritis in younger patients. Acta Orthop 2010; 81:161-4.

[6] Katz JN, Barrett J, Mahomed NN, Baron JA, Wright RJ, Losina E. Association between hospital and surgeon procedure volume and the outcomes of total knee replacement. J Bone Joint Surg Am 2004; 86A: 1909-16.

[7] Jacobs JJ, Urban RM, Hallab NJ, Skipor AK, Fischer A, Wimmer MA. Metal on-metal bearing surfaces. J Am Acad Orthop Surg 2009; 17:69-76.

[8] Knutsen G, Drogset JO, Engebretsen L, Grontvedt T, Isaksen V, Ludvigsen TC, et al. A randomized trial comparing autologous chondrocyte implantation with microfracture: findings at five years. J Bone Joint Surg Am 2007; 89:2105-12.

[9] Basad E, Ishaque B, Bachmann G, Sturz H, Steinmeyer J. Matrix-induced autologous chondrocyte implantation versus microfracture in the treatment of cartilage defects of the knee: a 2-year randomised study. Knee Surg Sports Traumatol Arthrosc 2010; 18:519-27.

[10] Agel J, et al. The burden of musculoskeletal conditions at the start of the new millennium. Geneva: World Health Organization; 2000.

[11] Rehman Q, Lane NE. Getting control of osteoarthritis pain: an update on treatment options. Postgrad Med 1999; 106:127-34.

[12] Murray CJL. The global burden of disease. Geneva: World Health Organization; 1996.

[13] Kenneth D.Osteoarthritis. Harrison's Principles of Internal Medicine 16th ed.2005.

[14] Romanes GJ. General introductions: Joints. In: Cunningham's Manual of practical anatomy. 15th ed. Vol (1). Oxford: Oxford University Press; 1986.

[15] Martini FH. Articulations: In: Fundamentals of anatomy and physiology. 4th ed. New Jersey: Prentice Hall; 1998.

[16] Snell RS. Basic anatomy. In: Clinical Anatomy. 7th ed. Place: Lippincott Williams & Wilkins; 2004

[17] Lim KKT, Shahid M and Sharif M. Recent advances in Osteoarthritis. Singapore Med J 1996; Vol 37: 189-193.

[18] Merskey H, Bogduk N. Classification of chronic pain: Description of chronic pain syndromes and definitions of pain terms. Seattle, WA: International Association for the Study of Pain Press; 1994.

[19] Sherrington CS. The integrative action of the nervous system. 2nd ed. New Haven: Yale University Press; 1947.

[20] Berne RM, Levy MN, editors. Physiology. 4th ed. St. Louis, MO: Mosby; 1998.

[21] Ganong WF. Review of medical physiology. 20th ed. Place: McGraw-Hill; 2001.

[22] Wyke B. The neurology of joints: a review of general principles. Clin Rheum Dis 1981; 57:233-9.

[23] Alvarez FJ, Fyffe RE. Nociceptors for the 21st century. Curr Rev Pain 2000; 4:451-8.

[24] Lawson SN. Phenotypes and function of somatic afferent nociceptive neurons with C, Adelta- or Aalpha/beta fibres. Exp Physiol 2002; 87:239-44.

[25] Wojtys EM, Beamann DN, Glover RA, et al. Innervation of the human knee joint by substance-P fibers. Arthroscopy 1990; 6:254-63.

[26] Kellgren JH, Samuel EP. The sensitivity and innervation of the articular capsule. J Bone Joint Surg Br 1950; 32:84-92.

[27] Dye SF, Chew MH. The use of scintigraphy to detect increased osseous metabolic activity about the knee. J Bone Joint Surg Am 1993; 75:1388-406.

[28] Creamer P, Hunts M, Dieppe P. Pain mechanisms in osteoarthritis of the knee: effect of intraarticular anesthetic. J Rheumatol 1996; 23:1031-36.

[29] Felson DT, Mclaughlin S, Goggins J, et al. Bone marrow edema and its relation to progression of knee osteoarthritis. Ann Intern Med 2003; 139:330-36

[30] Townes AS. Osteoarthritis. In: Barker LR, Burton JR, Zieve PD, editors. Principles of ambulatory medicine. 5th ed. Baltimore: Williams & Wilkins; 1999: 960-73.

[31] Jenkins BJ. Physiology of pain: Module 2 Lecture. In: PGCert/PGDip/MSc in Pain Management. Cardiff University, Wales; 1998.

[32] Farrel M, Gibson S, McMeeken J, et al. Pain and hyperalgesia in osteoarthritis of the hands. J Rheumatol 2000; 27:441-47

[33] Melzack R, et al. Central neuroplasticity and pathological pain. Ann NY Acad sci 2001; 933:157-74.

[34] Bonica JJ. History of pain concepts and pain therapy. Seminars in Anaesthesia 1985; 4:189-219.

[35] Brookoff D. Chronic pain: 1. A new disease? Hosp Pract 2000; 35:45-59.

[36] Brookoff D. Chronic pain: 1. A new disease? Hosp Pract 2000; 35:45-59.

[37] Bliss TV, Lømo T. Long-lasting potentiation of synaptic transmission in the dentate area of the anaesthetized rabbit following stimulation of the perforant path. J Physiol 1973; 232:331-56.

[38] Lømo T. The discovery of long-term potentiation. Philos Trans R Soc Lond B Biol Sci 2003; 358:617-20.

[39] Klein T, Magerl W, Hopf HC, et al. Perceptual correlates of nociceptive longterm potentiation and long-term depression in humans. J Neurosci 2004; 24:964-71.

[40] Sandkühler J. Learning and memory in pain pathways. Pain 2000; 88:113-8.

[41] Sandkühler J, Benrath J, Brechtel C, et al. Synaptic mechanisms of hyperalgesia. Prog Brain Res 2000; 129:81-100.

[42] Liu XG, Sandkühler J. Activation of spinal NMDA or neurokinin receptors induces long-term potentiation of spinal C-fibre-evoked potentials. Neurosci 1998; 86:1209-16.

[43] Rosenberg JM, Harrell C, Ristic H, et al. The effect of gabapentin on neuropathic pain. Clin J Pain 1997; 13:251-5.

Permissions

The contributors of this book come from diverse backgrounds, making this book a truly international effort. This book will bring forth new frontiers with its revolutionizing research information and detailed analysis of the nascent developments around the world.

We would like to thank Prof. Dhanjoo N. Ghista, for lending his expertise to make the book truly unique. He has played a crucial role in the development of this book. Without his invaluable contribution this book wouldn't have been possible. He has made vital efforts to compile up to date information on the varied aspects of this subject to make this book a valuable addition to the collection of many professionals and students.

This book was conceptualized with the vision of imparting up-to-date information and advanced data in this field. To ensure the same, a matchless editorial board was set up. Every individual on the board went through rigorous rounds of assessment to prove their worth. After which they invested a large part of their time researching and compiling the most relevant data for our readers. Conferences and sessions were held from time to time between the editorial board and the contributing authors to present the data in the most comprehensible form. The editorial team has worked tirelessly to provide valuable and valid information to help people across the globe.

Every chapter published in this book has been scrutinized by our experts. Their significance has been extensively debated. The topics covered herein carry significant findings which will fuel the growth of the discipline. They may even be implemented as practical applications or may be referred to as a beginning point for another development. Chapters in this book were first published by InTech; hereby published with permission under the Creative Commons Attribution License or equivalent.

The editorial board has been involved in producing this book since its inception. They have spent rigorous hours researching and exploring the diverse topics which have resulted in the successful publishing of this book. They have passed on their knowledge of decades through this book. To expedite this challenging task, the publisher supported the team at every step. A small team of assistant editors was also appointed to further simplify the editing procedure and attain best results for the readers.

Our editorial team has been hand-picked from every corner of the world. Their multi-ethnicity adds dynamic inputs to the discussions which result in innovative outcomes. These outcomes are then further discussed with the researchers and contributors who give their valuable feedback and opinion regarding the same. The feedback is then collaborated with the researches and they are edited in a comprehensive manner to aid the understanding of the subject.

Apart from the editorial board, the designing team has also invested a significant amount of their time in understanding the subject and creating the most relevant covers. They scrutinized every image to scout for the most suitable representation of the subject and create an appropriate cover for the book.

The publishing team has been involved in this book since its early stages. They were actively engaged in every process, be it collecting the data, connecting with the contributors or procuring relevant information. The team has been an ardent support to the editorial, designing and production team. Their endless efforts to recruit the best for this project, has resulted in the accomplishment of this book. They are a veteran in the field of academics and their pool of knowledge is as vast as their experience in printing. Their expertise and guidance has proved useful at every step. Their uncompromising quality standards have made this book an exceptional effort. Their encouragement from time to time has been an inspiration for everyone.

The publisher and the editorial board hope that this book will prove to be a valuable piece of knowledge for researchers, students, practitioners and scholars across the globe.

List of Contributors

Dhanjoo N. Ghista
Department of Graduate and Continuing Education, Framingham State University, Framingham, Massachusetts, USA

John I. Anetor, Segun Adeola and Ijeoma Esiaba
Department of Chemical Pathology, College of Medicine, University of Ibadan, Nigeria

Gloria O. Anetor
Department of Human Kinetics and Health Education, Faculty of Education, University of Ibadan, Nigeria

John T. Hancock
University of the West of England, Bristol, UK

Hidenori Koyama and Tetsuya Yamamoto
Department of Internal Medicine, Division of Endocrinology and Metabolism, Hyogo College of Medicine, Japan

Magdalena Labieniec-Watala, Karolina Siewier and Slawomir Gierszewski
University of Lodz, Department of Thermo biology, Poland

Cezary Watala
Medical University of Lodz, Department of Hemostasis and Hemostatic Disorders, Poland

Ahmed Morsy Ahmed
Faculty of Agriculture, Ain shams University, Egypt

O. O. Oguntibeju, A. J. Esterhuyse and E. J. Truter
Oxidative Stress Research Centre, Department of Biomedical Sciences, Faculty of Health & Wellness Sciences, Cape Peninsula University of Technology, South Africa

Scott Cooper, Melissa Dick, Alexander Emmott, Paul Jonak, Léonie Rouleau and Richard L. Leask
Department of Chemical Engineering, McGill University, Canada

Kingsley Enohumah
The Rotunda Hospital, Dublin, Ireland

Printed in the USA
CPSIA information can be obtained
at www.ICGtesting.com
JSHW011419221024
72173JS00004B/589